AIRCRAFT SHEET METAL

By Nick Bonacci

Library of Congress Cataloging-in-publication number: 92-24394

JS312633B

Table Of Contents

PREFACE

This book on *Aircraft Sheet Metal* is one of a series of specialized training manuals prepared for aviation maintenance personnel. This series is part of a programmed learning course produced by Jeppesen Sanderson, one of the largest suppliers of aviation maintenance training materials in the world. This program is part of a continuing effort to improve the quality of education for aviation mechanics throughout the world.

The purpose of each Jeppesen Sanderson training series is to provide basic information on the operation and principles of the various aircraft systems and their components. Specific information on exact application procedures of a product should be obtained from the manufacturer through his appropriate maintenance manuals and followed in detail for the best results.

This particular manual on *Aircraft Sheet Metal* includes a series of carefully prepared questions and answers to emphasize key elements of the study, and to encourage you to continually test yourself for accuracy and retention as you use this book.

Some of the terminology used in this book may be new to you. Throughout the text, you will find terms that are defined in the Glossary at the back of the book highlighted as follows: ***glossary item***.

Acknowledgements
The validity of any program such as this is enhanced immeasurably by the cooperation shown by recognized experts in the field, and by the willingness of the various manufactuers to share their literature and answer countless questions in the preparation of these programs.

We would like to mention, especially, our appreciation for help given us by:

Bill DeWalt, ALCOA Rivets
John Weaver, ALCOA
Ray Pedersen, Boeing Commercial Aircraft Company
James Schubert, Boeing Commercial Aircraft Company
Mike Potts, Beechcraft
H. Dean Humphrey, Cessna
Ronald Folkerts, Greenlee Tool Company
Michael Schuster, Hi-Shear Corporation
Ramon L. Hurd, Huck Aerospace Fasteners
Ed Mares, P.B. (BRILES) Fasteners
Roy Moosa, Cherry Aerospace Fasteners
Joe Wetstein, Professor of Aviation — Lewis University
Dr. Bette Bonacci, Professor of English — Lewis University
Dr. John Bonacci, Structural Engineer

For product, service, or sales information call **1-800-621-JEPP, 303-799-9090, or FAX 303-784-4153**. If you have comments, questions, or need explanations about any component of our Maintenance Training System, we are prepared to offer assistance at any time. If your dealer does not have a Jeppesen catalog, please request one and we will promptly send it to you. Just call the above telephone number, or write:

Marketing Manager, Training Products
Jeppesen Sanderson, Inc.
55 Inverness Drive East
Englewood, CO 80112-5498

Please direct inquiries from Europe, Africa, and the Middle East to:

Jeppesen & Co., GmbH
P. O. Box 70-05-51
Walter-Kolb-Strasse 13
60594 Frankfurt
GERMANY
Tel: 011-49-69-961240
Fax: 011-49-69-96124898

Chapter I
Aircraft Structural Components

The major aircraft structures are wings, fuselage, and empennage. The primary flight control surfaces, located on the wings and empennage, are ailerons, elevators, and rudder. These parts are connected by seams, called joints.

A. Joints

All joints constructed using rivets, bolts, or special fasteners are ***lap joints.*** Fasteners cannot be used on joints in which the materials to be joined do not overlap — for example, butt, tee and edge joints. A ***fayed edge*** (Figure 1-1) is a type of lap joint made when two metal surfaces are butted up against one another in such a way as to overlap.

Internal aircraft parts are manufactured in four ways: Milling, stamping, ***bending,*** and extruding. The metal of a milled part is transformed from cast

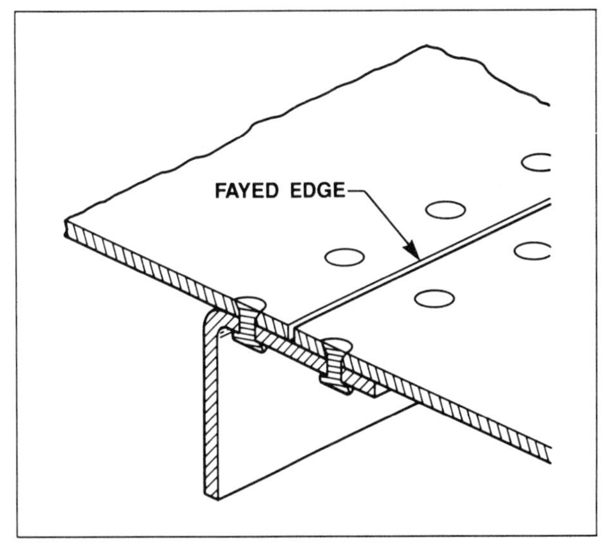

Fig. 1-1 Fayed edge joint.

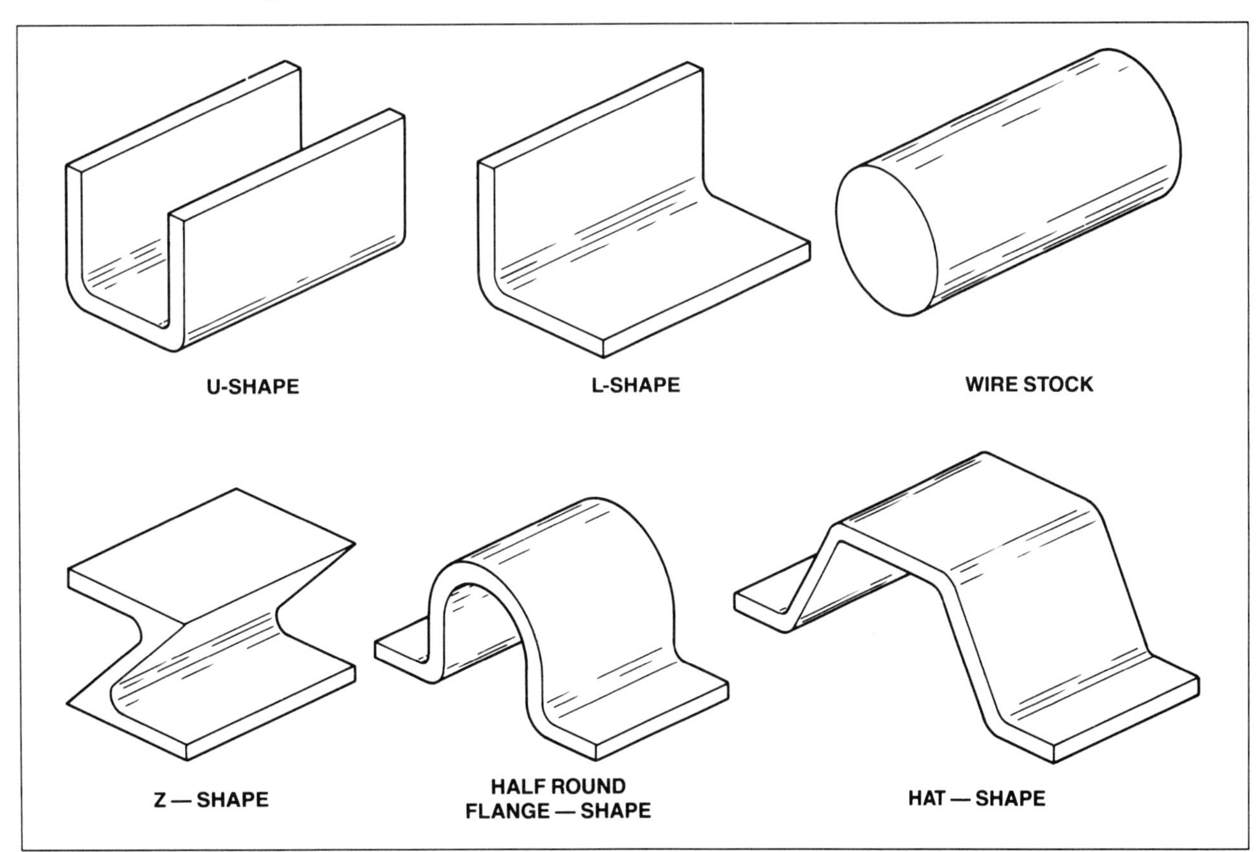

Fig. 1-2 Extruded Shapes.

to wrought by first shaping and then either chemically etching or grinding it.

A stamped part is ***annealed,*** placed in a forming press, and then re-heat treated.

Bent parts are made by sheet metal mechanics using the ***bend allowance*** and layout procedures.

An ***extrusion*** is an aircraft part which is formed by forcing metal through a preshaped die. The resulting wrought forms are used as ***spars, stringers, longerons,*** or ***channels.***

Figure 1-2 shows a selection of extruded shapes used for the construction of many aircraft internal parts. In order for metal to be extruded, bent, or formed, it must first be made ***malleable*** and ***ductile*** by annealing. After the forming operation, the metal is re-heat treated and age hardened.

B. Wings

1. General

The aircraft wing has to be strong enough to withstand the positive forces of flight as well as the negative forces of landing. Metal wings are of two types: Semicantilever and full ***cantilever.*** Semicantilever, or braced, wings are used on light aircraft. They are externally supported by struts or flying wires which connect the wing spar to the fuselage.

A full cantilever wing (Figure 1-3) is usually made of stronger metal. It requires no external bracing or support. The skin carries part of the wing ***stress.*** Parts common to both wing designs are spars, ***compression ribs,*** former ribs, stringers, stress plates, gussets, wing tips and wing skins (Figure 1-4).

2. Spars

Two or more spars are used in the construction of a wing. They carry the main longitudinal — butt to tip — load of the wing. Both the spar and a compression rib connect the wing to the fuselage.

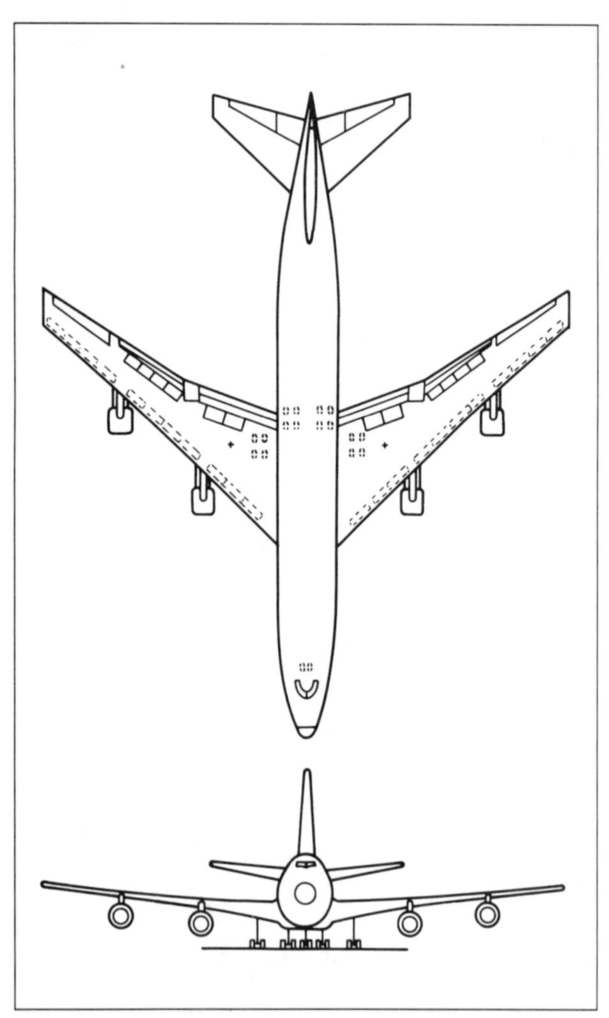

Fig. 1-3 Full cantilever wing — Boeing 747.

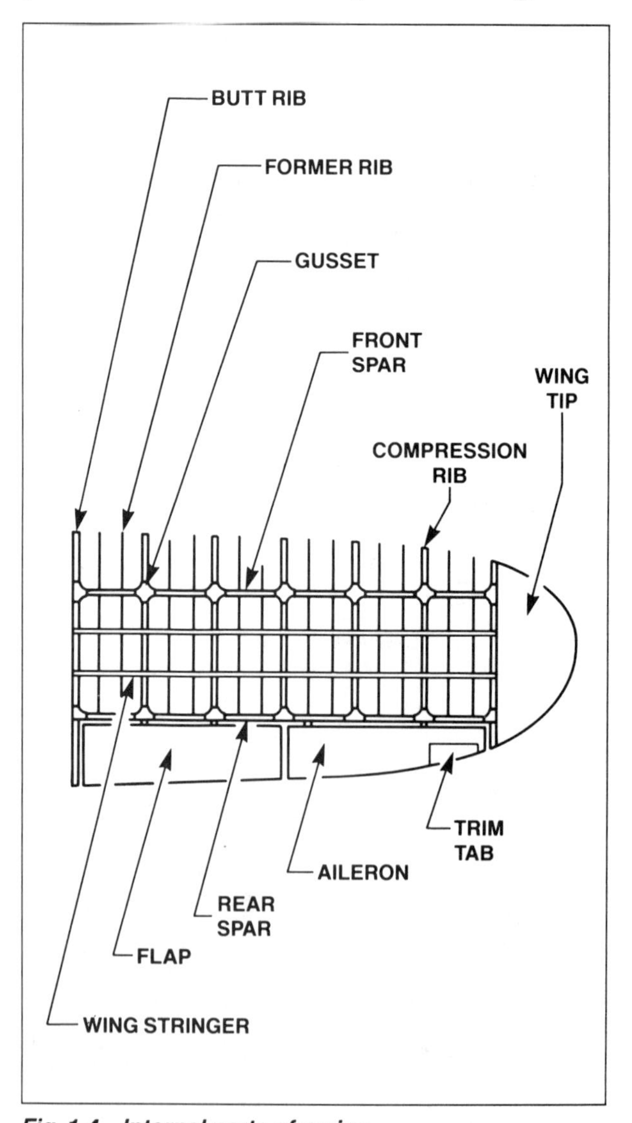

Fig. 1-4 Internal parts of a wing.

3. Compression Ribs

Compression ribs carry the main load in the direction of flight, from leading edge to trailing edge. On some aircraft the compression rib is a structural piece of tubing separating two main spars. The main function of the compression rib is to absorb the force applied to the spar when the aircraft is in flight.

4. Former Ribs

A former rib, which is made from light metal, attaches to the stringers and wing skins to give the wing its aerodynamic shape. Former ribs can be classified as nose ribs, trailing edge ribs, and mid ribs running fore and aft between the front and rear spar on the wing. ***Formers*** are not considered primary structural members.

5. Stringers

Stringers are made of thin sheets of preformed extruded or hand-formed ***aluminum alloy.*** They run front to back along the fuselage and from wing butt to wing tip. Riveting the wing skin to both the stringer and the ribs gives the wing additional strength.

6. Stress Plates

Stress plates are used on wings to support the weight of the fuel tank. Some stress plates are made of thick metal and some are of thin metal corrugated for strength. Stress plates are usually held in place by long rows of machine screws, with self-locking nuts, that thread into specially mounted channels. The stress-plate channeling is riveted to the spars and compression ribs.

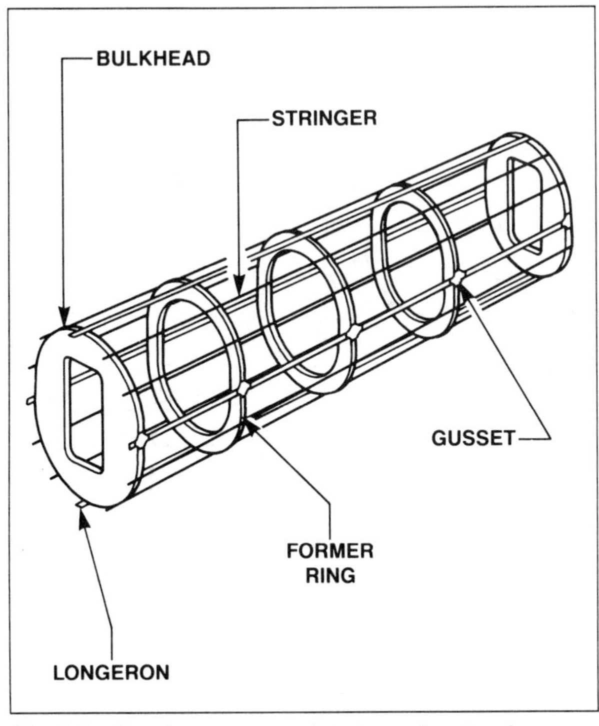

Fig. 1-5 Semimonocoque fuselage structural components.

7. Gussets

Gussets, or gusset plates, are used on aircraft to join and reinforce intersecting structural members. Gussets are used to transfer stresses from one member to another at the point where the members join.

8. Wing Tips

The wing tip, the outboard end of the wing, has two purposes: To aerodynamically smooth out the wing tip air flow and to give the wing a finished look.

9. Wing Skins

Wing skins cover the internal parts and provide for a smooth air flow over the surface of the wing. On full cantilever wings, the skins carry stress. However, all wing skins are to be treated as primary structures whether they are on braced or full cantilever surfaces.

C. Fuselage Assemblies

1. General

There are two types of metal aircraft fuselages: Full ***monocoque*** and ***semimonocoque.*** The full monocoque fuselage has fewer internal parts and a more highly stressed skin than the semimonocoque fuselage, which uses internal bracing to obtain its strength.

The full monocoque fuselage is generally used on smaller aircraft, because the stressed skin eliminates the need for stringers, former rings, and other types of internal bracing, thus lightening the aircraft structure.

The semimonocoque fuselage derives its strength from the following internal parts: Bulkheads, longerons, keel beams, drag struts, body supports, former rings, and stringers (Figure 1-5).

2. Bulkheads

A bulkhead is a structural partition, usually located in the fuselage, which normally runs perpendicular to the keel beam or longerons. A few examples of bulkhead locations are where the wing spars connect into the fuselage, where the cabin pressurization domes are secured to the fuselage structure, and at cockpit passenger or cargo entry doors.

3. Longerons And Keel Beams

Longerons and keel beams perform the same function in an aircraft fuselage. They both carry the bulk of the load traveling fore and aft. The keel beam and longerons, the strongest sections of the airframe, tie its weight to other aircraft parts, such as powerplants, fuel cells, and the landing gears.

4. Drag Struts And Other Fittings

Drag struts and body support fittings are other primary structural members. Drag struts are used on large jet aircraft to tie the wing to the fuselage center section. Body support fittings are used to support the structures which make up bulkhead or floor truss sections.

Former rings and fuselage stringers are not primary structural members. Former rings are used to give shape to the fuselage. Fuselage stringers running fore and aft are used to tie in the bulkheads and former rings.

D. Empennage Section

1. General

The ***empennage*** is the tail section of an aircraft. It consists of a horizontal stabilizer, elevator, vertical stabilizer and rudder (Figure 1-6). The conventional empennage section contains the same kind of parts used in the construction of a wing. The internal parts of the stabilizers and their flight controls are made with spars, ribs, stringers and skins. Also, tail sections, like wings, can be externally or internally braced.

2. Horizontal Stabilizer And Elevator

The horizontal stabilizer is connected to a primary control surface, i.e., the elevator. The elevator causes the nose of the aircraft to pitch up or down. Together, the horizontal stablizer and elevator provide stability about the horizontal axis of the aircraft. On some aircraft the horizontal stabilizer is made movable by a ***screw-jack assembly*** which allows the pilot to trim the aircraft during flight.

3. Vertical Stabilizer And Rudder

The vertical stabilizer is connected to the aft end of the fuselage and gives the aircraft stability about the vertical axis. Connected to the vertical stabilizer is the rudder, the purpose of which is to turn the aircraft about its vertical axis.

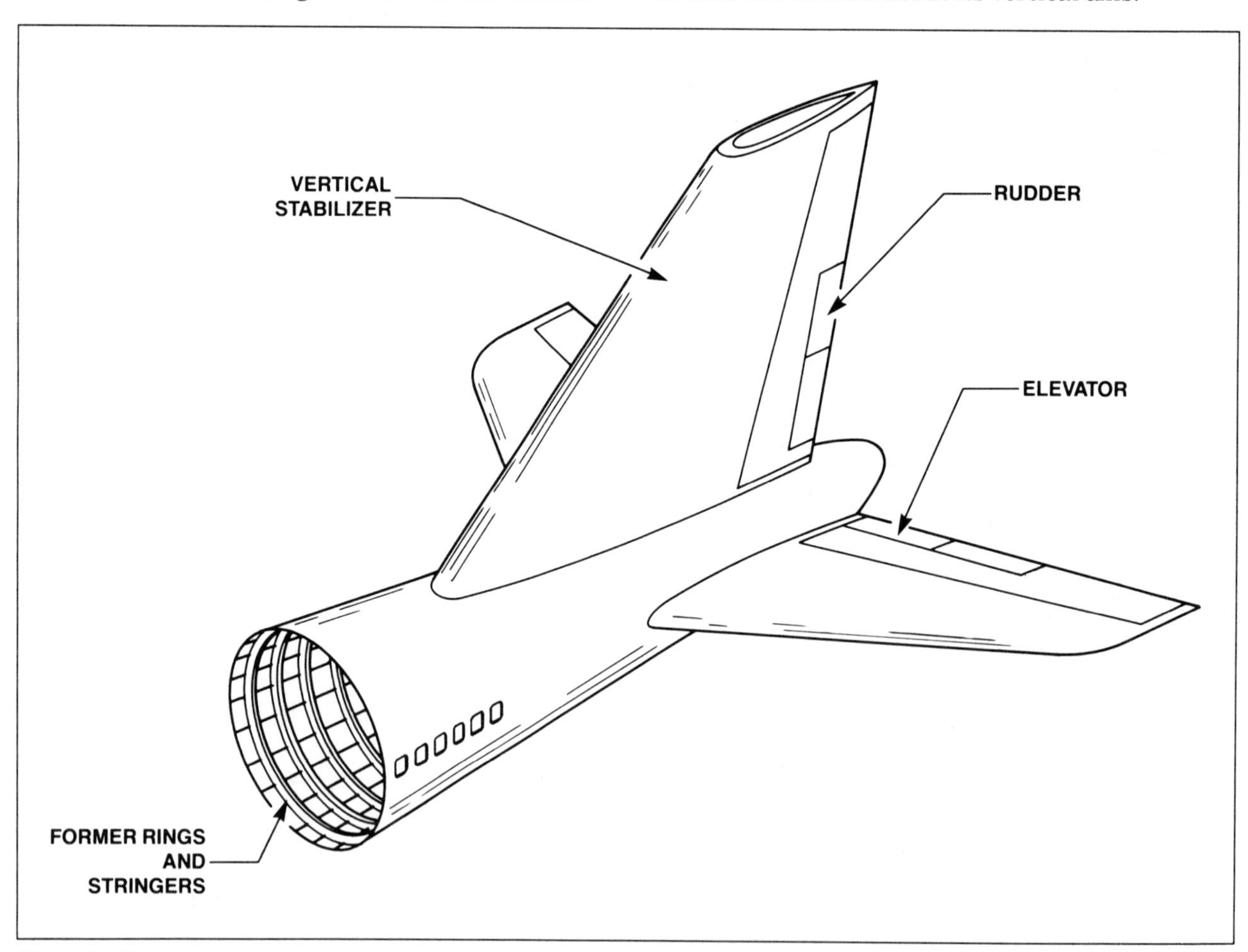

Fig. 1-6 Empennage.

E. Flight Control

1. Ailerons

Elevators and rudders are primary flight controls in the tail section. Ailerons are primary flight controls connected to the wings. Located on the outboard portion of the wing, they allow the aircraft to turn about the longitudinal axis.

When the right aileron is moved upward, the left one goes down, thus causing the aircraft to roll to the right. Because this action creates a tremendous force, the ailerons must be constructed in such a way as to withstand it.

Flight controls other than the three primary ones are needed on high-performance aircraft. On the wings of a wide-body jet, for example, there are as many as thirteen flight controls, including high- and low-speed ailerons, flaps, and spoilers.

2. Flaps And Spoilers

Wing flaps increase the lift for take-off and landing. Inboard and outboard flaps, on the trailing edge of the wing, travel from full up, which is neutral aerodynamic flow position, to full down, causing air to pile up and create lift. Leading edge flaps — ***Krueger flaps*** and ***variable-camber flaps*** (Figure 1-7) — increase the wing chord size and thus allow the aircraft to take off or land on a shorter runway.

Spoilers, located in the center section span-wise, serve two purposes. They assist the high-speed ailerons in turning the aircraft during flight, and they are used to kill the aerodynamic lift during landing by spreading open on touchdown.

3. Trim Tabs

Connected to the primary flight controls are devices called trim tabs. They are used to make fine adjustments to the flight path of an aircraft. Trim tabs are constructed like wings or ailerons, but are considerably smaller.

QUESTIONS:

1. *Name the primary sections of a modern aircraft.*
2. *Name the primary control surfaces of an aircraft.*
3. *List five auxiliary flight controls.*
4. *What is a semicantilever wing?*
5. *What is a full monocoque fuselage?*
6. *A spar is to a longeron as a compression rib is to a ____________________.*
7. *What is the main fore-and-aft structural member on an aircraft?*
8. *Name the parts that make up an aircraft empennage section.*
9. *What happens to the upper skin areas of a wing of an aircraft in flight?*
10. *Name two types of fuselage constructions.*
11. *Why are wing spoilers used on modern jetliners?*
12. *What is the main difference between a braced and a full cantilever wing?*
13. *What are the primary tip-to-butt load-carrying members of a wing?*
14. *Name three internal parts of an aileron.*
15. *Why are some control surface skin coverings corrugated?*

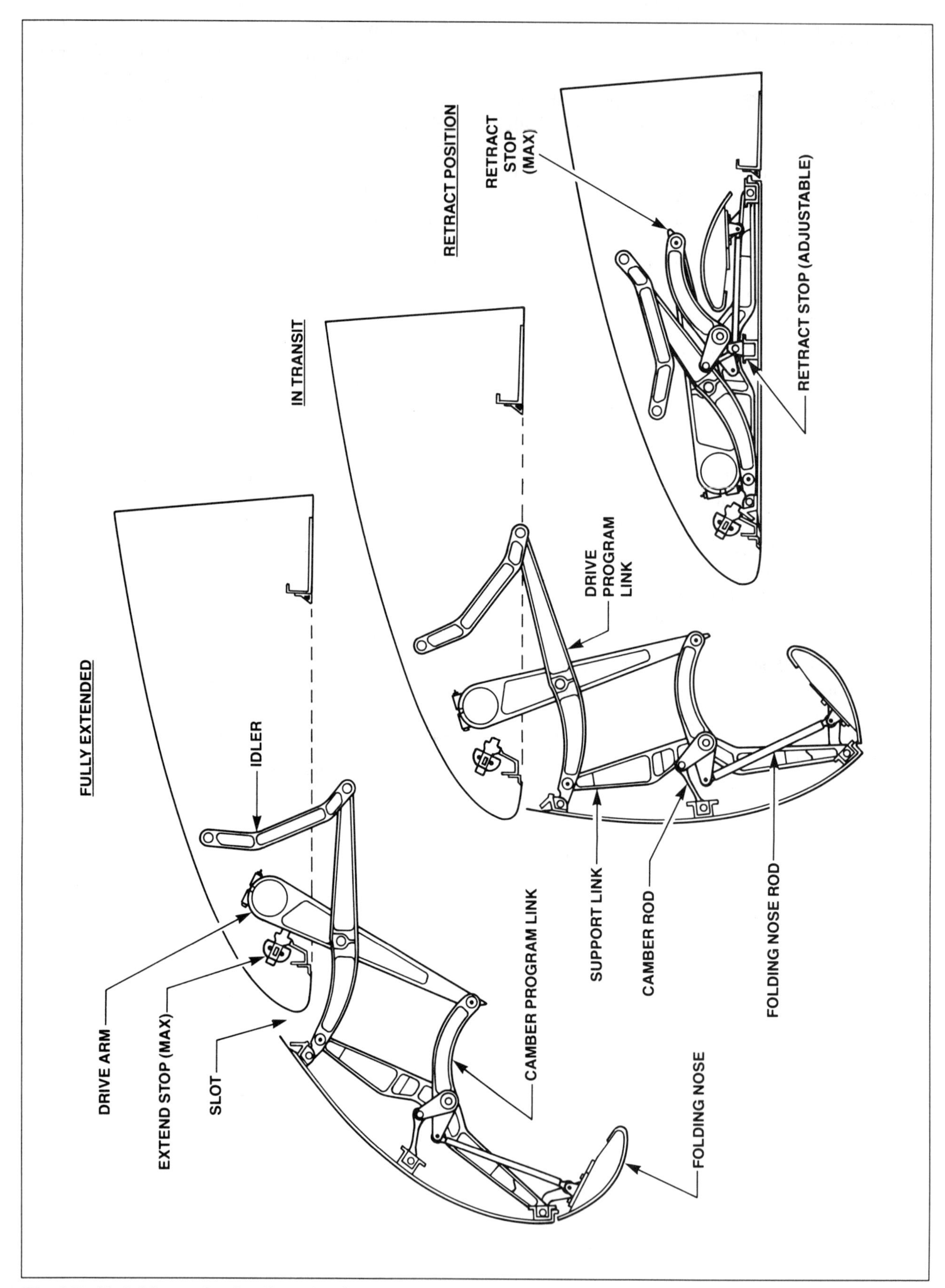

Fig. 1-7 Variable camber leading edge flaps schematic.

Chapter II
Aircraft Repair Tools

A. Hand Tools

1. Measuring Scale

A tool used by sheet metal mechanics for measuring a layout is the six-inch scale (Figure 2-1). It has a decimal equivalent chart on one side and 1/64-inch increments on the other.

2. Motor And Drill

Two important hand tools are the motor and ***drill.*** The motor, which drives the drill, can be pneumatic (air-powered) or electrical. In aircraft repair work, the air-driven motor is usually preferred because electrical lines can be accidentally cut while working in close quarters on an aircraft, and because all air motors have variable drill speeds and are more durable than most electrical motors. However, electrical motors are less expensive, and some of the newer models are battery operated.

Drills are used to cut holes in aluminum skin for rivets, special fasteners, and bolts.

A precision tool, the drill requires special care. Always store it in a plastic container to prevent damaging the tip and reamers.

Drills are cataloged in four ways: Fraction, decimal, letter (A to Z), number (1 to 80), and metric. The number drills most commonly used in the aircraft industry are #40, #30, #21, and #11. These drills are slightly larger than the rivet diameters which correspond to them (Figure 2-2).

The difference between the rivet diameter and the recommended drill size, .002 to .004 of an inch, allows the rivet to be inserted easily and also provides for shank expansion.

3. Drill Description

A drill has three main parts: the tip, the body, and the shank (Figure 2-4).

The tip of the drill is made by the webbing, which looks like an hourglass when viewed from the tip or point.

The body of the drill is composed of the flutes and spiral webbing. The flutes are the valleys between the spiral webbing of the drill. They aid in cooling the tip, provide a path for lubricating the tip, and allow the drill cuttings to be removed.

The drill shank is the part of the drill that is clamped into place by the motor's chuck. The drill

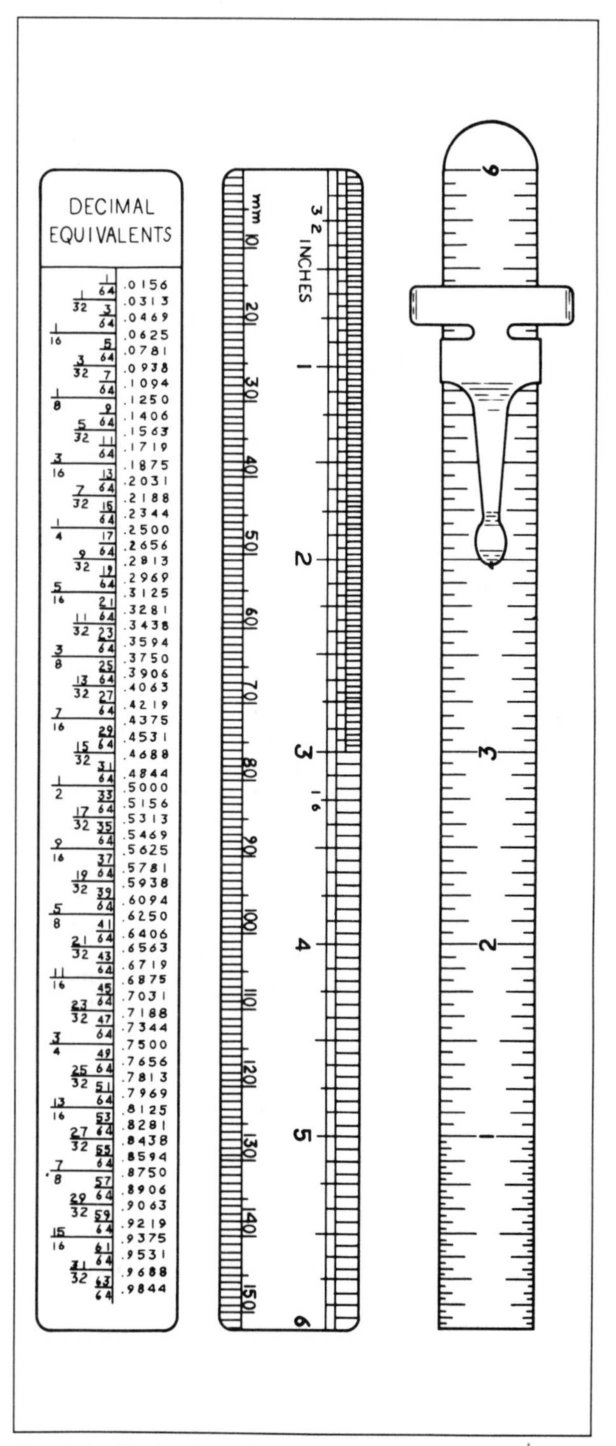

Fig. 2-1 Six-inch scales.

DRILL SIZE	RIVET DIAMETER	DIFFERENCE
#40 = .098	3/32 = .0937	.0043
#30 = .1285	1/8 = .125	.0035
#21 = .159	5/32 = .15625	.00275
#11 = .191	3/16 = .1875	.0035

Fig. 2-2 Rivet sizes and drill sizes.

Size	Decimal equivalent	Size	Decimal equivalent	Size	Decimal equivalent
1/2	0.5000	3	0.2130	3/32	0.0937
31/64	0.4844	4	0.2090	42	0.0935
15/32	0.4687	5	0.2055	43	0.0890
29/64	0.4531	6	0.2040	44	0.0860
7/16	0.4375	13/64	0.2031	45	0.0820
27/64	0.4219	7	0.2010	46	0.0810
Z	0.4130	8	0.1990	47	0.0785
13/32	0.4062	9	0.1960	5/64	0.0781
Y	0.4040	10	0.1935	48	0.0760
X	0.3970	11	0.1910	49	0.0730
25/64	0.3906	12	0.1890	50	0.0700
W	0.3860	3/16	0.1875	51	0.0670
V	0.3770	13	0.1850	52	0.0635
3/8	0.3750	14	0.1820	1/16	0.0625
U	0.3680	15	0.1800	53	0.0595
23/64	0.3594	16	0.1770	54	0.0550
T	0.3580	17	0.1730	55	0.0520
S	0.3480	11/64	0.1719	3/64	0.0469
11/32	0.3437	18	0.1695	56	0.0465
R	0.3390	19	0.1660	57	0.0430
Q	0.3320	20	0.1610	58	0.0420
21/64	0.3281	21	0.1590	59	0.0410
P	0.3230	22	0.1570	60	0.0400
O	0.3160	5/32	0.1562	61	0.0390
5/16	0.3125	23	0.1540	62	0.0380
N	0.3020	24	0.1520	63	0.0370
19/64	0.2969	25	0.1495	64	0.0360
M	0.2950	26	0.1470	65	0.0350
L	0.2900	27	0.1440	66	0.0330
9/32	0.2812	9/64	0.1406	1/32	0.0320
K	0.2810	28	0.1405	67	0.0312
J	0.2770	29	0.1360	68	0.0310
I	0.2720	30	0.1285	69	0.0292
H	0.2660	1/8	0.1250	70	0.0280
17/64	0.2656	31	0.1200	71	0.0260
G	0.2610	32	0.1160	72	0.0250
F	0.2570	33	0.1130	73	0.0240
E-1/4	0.2500	34	0.1110	74	0.0225
D	0.2460	35	0.1100	75	0.0210
C	0.2420	7/64	0.1094	76	0.0200
B	0.2380	36	0.1065	77	0.0180
15/64	0.2344	37	0.1040	1/64	0.0160
A	0.2340	38	0.1015	78	0.0156
1	0.2280	39	0.0995	79	0.0145
2	0.2210	40	0.0980	80	0.0135
7/32	0.2187	41	0.0960		

Fig. 2-3 Twist drill sizes.

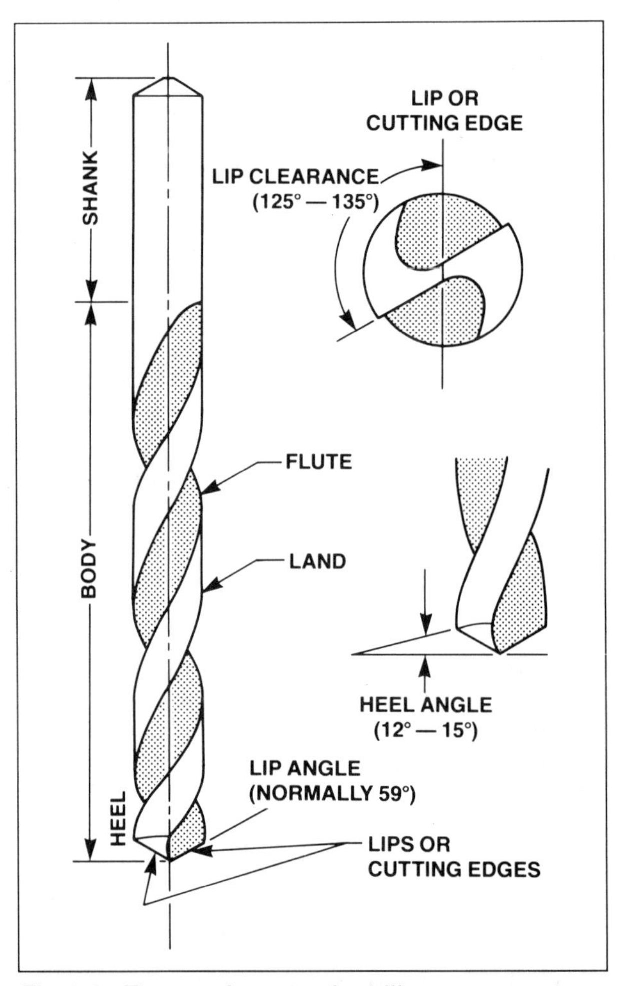

Fig. 2-4 Three main parts of a drill.

size is stamped on the end of its shank. A chuck key is used to tighten the drill into the chuck.

A drill must be able to cut holes that are round, straight, and free from cracks. The drill must be sharpened to an angle determined by the type of metal being drilled. The drill speed must also be adjusted for different hardness of metal.

Drills used to cut holes in aircraft aluminum should have an ***included angle*** of 118°, a ***sharpening angle*** of 59°, and high tip speed. They should be operated with steady, downward hand pressure.

Drills used to open holes in stainless steel should have an included angle of 140°, a sharpening angle of 70° and slow tip speed. Heavy pressure on the drill is necessary.

Soft materials, such as lead, copper, brass, soft aluminum and Plexiglas, require an included angle of 90°, a sharpening angle of 45°, and a speed which varies with the material being drilled.

The included angle is the tip angle (drill point) as viewed from the center line of the drill body. The sharpening angle is one half of the included angle, because when a drill is sharpened, only one lip can come in contact with the grinding wheel. Figure 2-5 shows the included and sharpening angles.

A drill-sharpening gauge is used to check the sharpening angle of a drill. Figure 2-6 shows how to make the gauge. During sharpening, the drill tip must be kept cool at all times by dipping it frequently into cold water. If it overheats, it will lose its temper and not retain its sharp cutting edge.

4. Deburring Tool

If burrs, slivers which form at the edges of holes whenever a drill breaks through the metal, are not removed, they will deform the joint by creating a bulge between the skins when the fasteners are installed into the holes. A good ***deburring*** tool is a twist drill one or two times larger than the hole. Insert it into the hole and spin by hand until all the burrs are removed.

5. Microstop Countersink Cutters

There are three types of ***countersink*** cutters: A non-adjustable fixed cutter, a ***stop countersink*** which is equipped with a rivet hole pilot independent

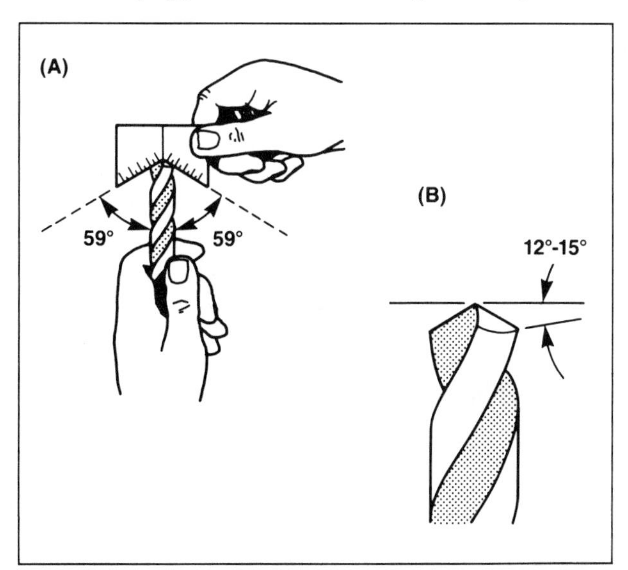

Fig. 2-5 Included and sharpening angles of a drill.

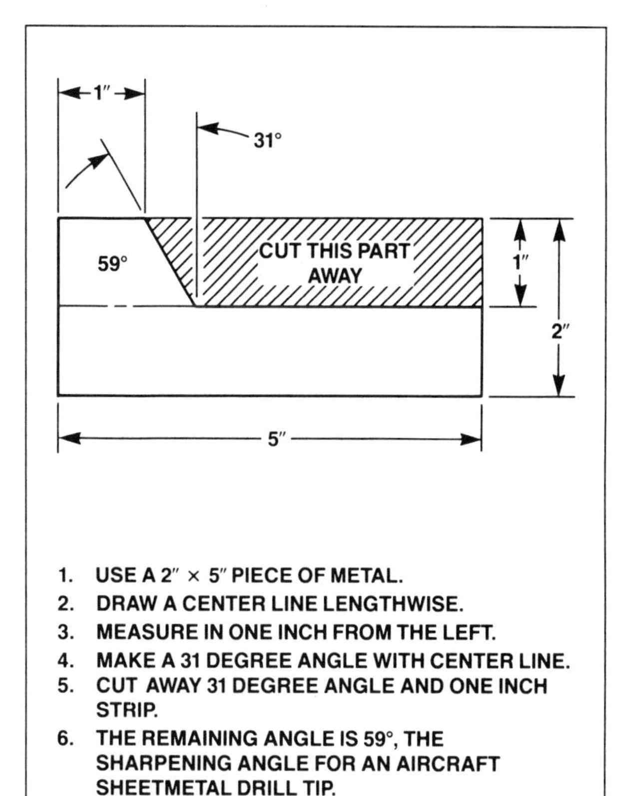

Fig. 2-6 Drill-sharpening gauge.

of the cutter, and a microstop countersink cutter which includes a hole pilot.

The microstop countersink cutter (Figure 2-7) allows the user to make micro adjustments when setting the depth of the countersink cutter. Another advantage of the microstop countersink is that it is more durable than the other types.

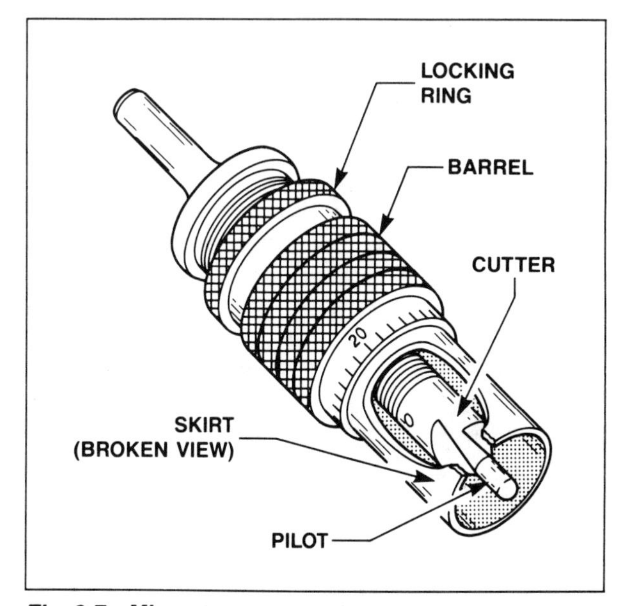

Fig. 2-7 Microstop countersink.

6. Cleco Clamps And Pliers

Cleco clamps are used to align parts prior to being reriveted to an aircraft. The clamps are installed with Cleco pliers (Figure 2-8). The color of the Cleco clamp indicates the diameter of the rivet it is to be used with. Four commonly used sizes are 3/32 of an inch (silver), 1/8 of an inch (copper), 5/32 of an inch (black), and 3/16 of an inch (brass, gold).

7. Rivet Guns

The hand tool commonly used to drive rivets is a rivet gun (Figure 2-9). Rivet guns are powered by compressed air and are classified as light-, medium-, or heavy-hitting. A light-hitting gun is used to install 1/32-inch and 1/8-inch diameter rivets. Medium-hitting guns are used to install 5/32-inch and 3/16-inch diameter rivets. Heavy-hitting guns are used to install larger diameter rivets and some special fasteners.

8. Rivet Sets Or Gun Sets

There are two types of gun sets: The universal and the countersink (Figure 2-10). The universal gun set is sized to fit on the driven end of the various shapes of rivet heads. The opposite end of the universal gun set fits into the rivet gun barrel and is held in place by a ***beehive retainer spring.***

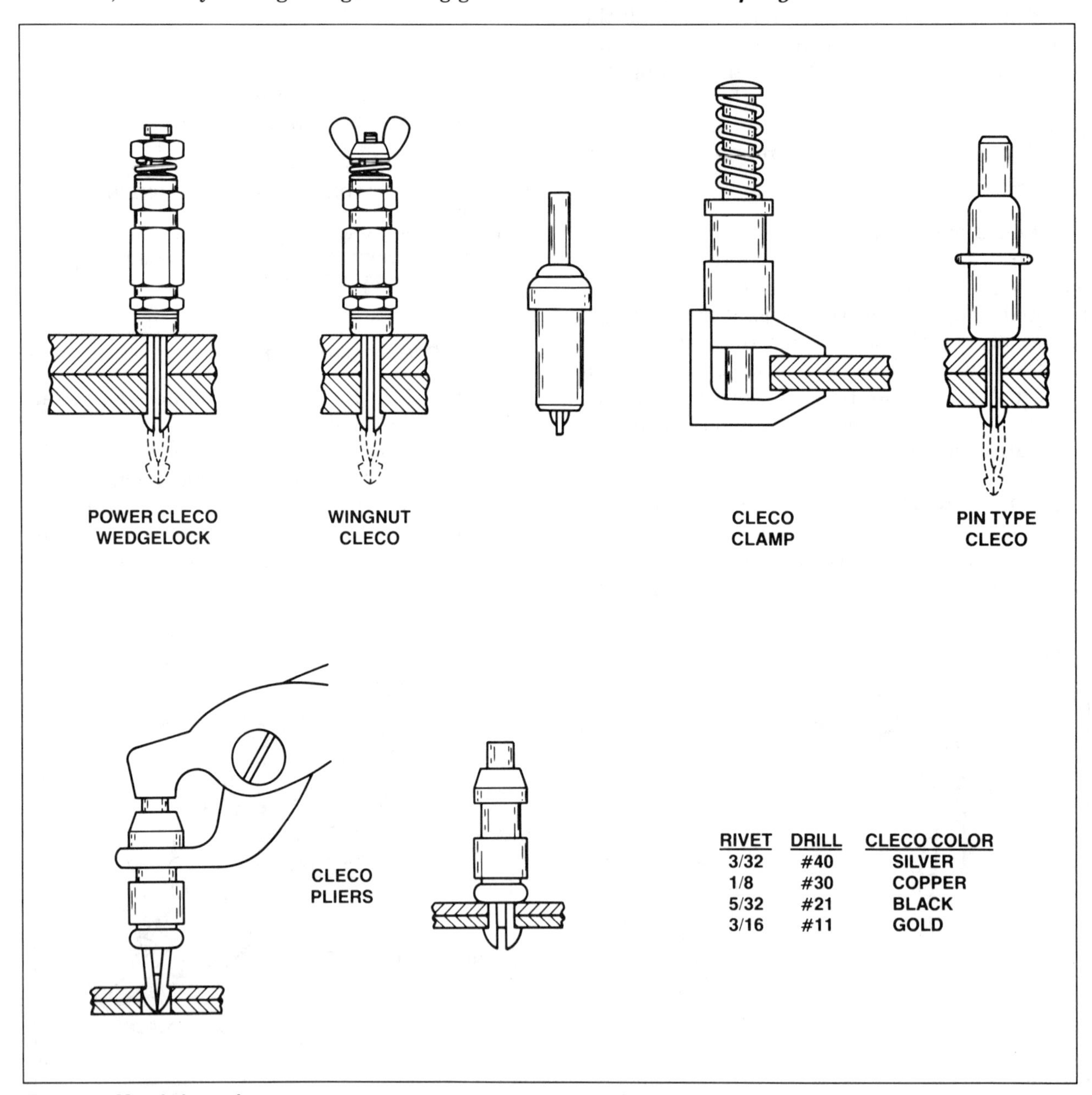

Fig. 2-8 Metal cleco clamps.

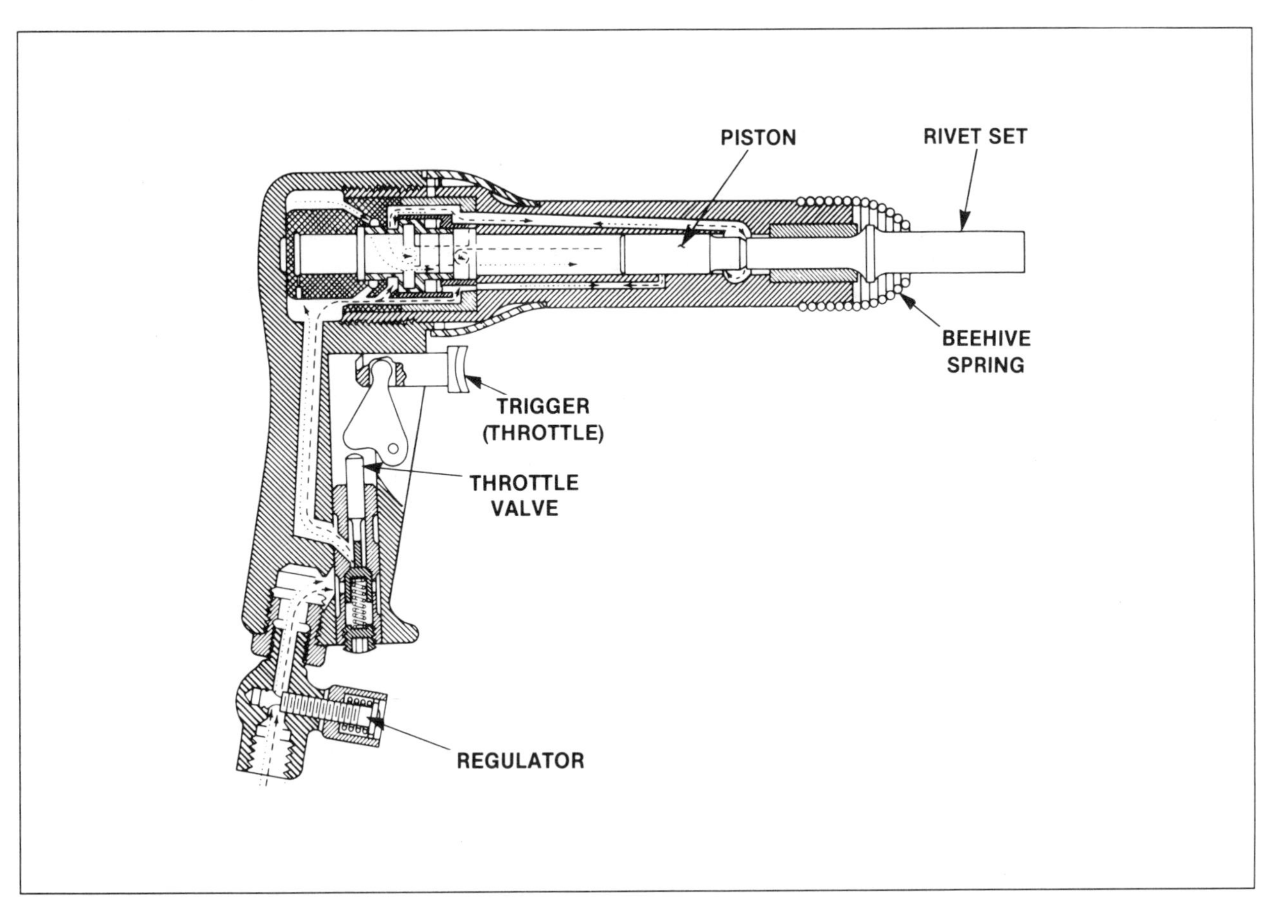

Fig. 2-9 Rivet gun.

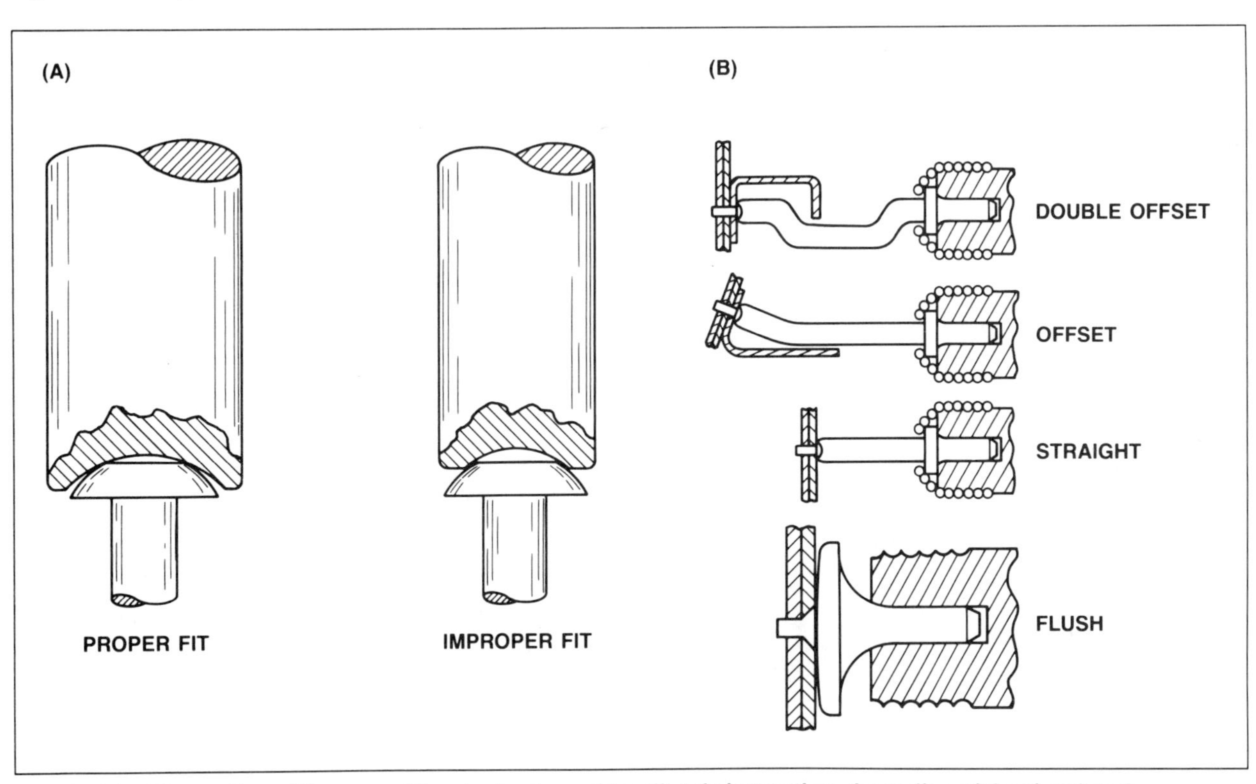

***Fig. 2-10A** The radius of the cup of the rivet set must be slightly larger than the radius of the rivet head.*
***Fig. 2-10B** Types of rivet sets.*

The countersink gun set is made to fit on the driven end of any size flush head rivet. A beehive retainer spring will not fit over the gun set retainer ring of the countersink gun set. It uses a specially constructed retainer spring. Figure 2-11 pictures both types of retainer springs. The retainer ring, located in the middle of the gun set, holds the retainer spring as it is threaded onto the gun barrel and prevents the gun set from accidentally flying off. The rivet gun should never be fired if the gun set is not held in place by the retaining spring.

9. Rivet Cutters

Rivet cutters are used to cut rivets to size prior to driving. The rivet cutter has a stack of thickness gauges which are used to determine the correct rivet length by measuring the space between the rivet head and the cutting edge (Figure 2-12).

When rivet cutters are not available, the rivets can be cut to size using a pair of diagonal cutting pliers. The rivet is cut by squeezing together the two rotating plates connected to the cutter handles.

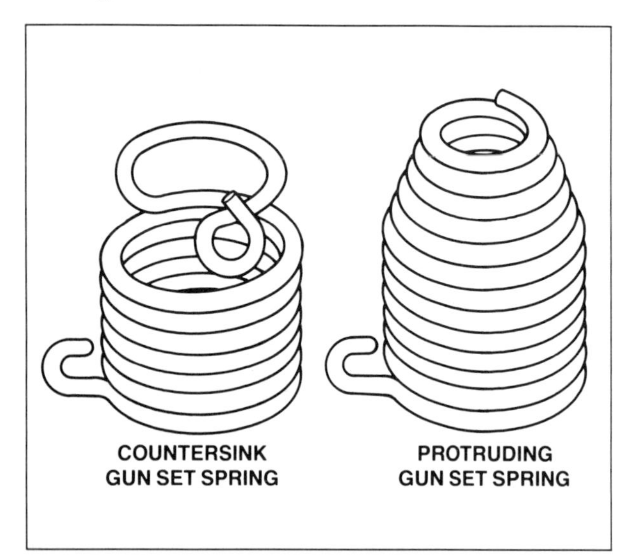

Fig. 2-11 Gun set retainer springs.

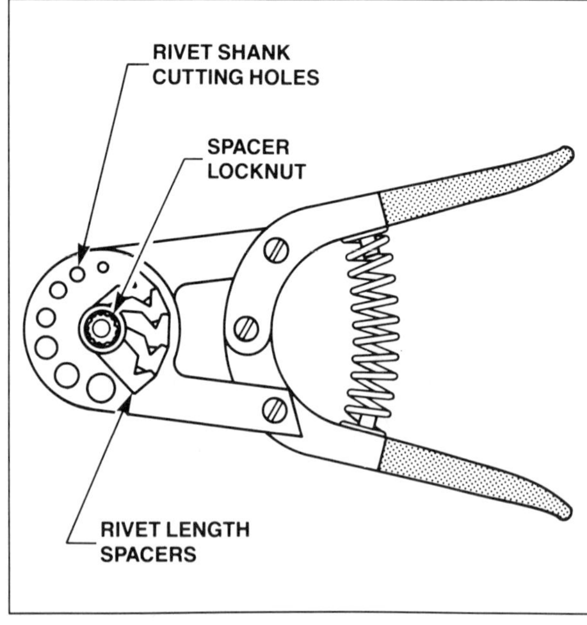

Fig. 2-12 Rivet cutters.

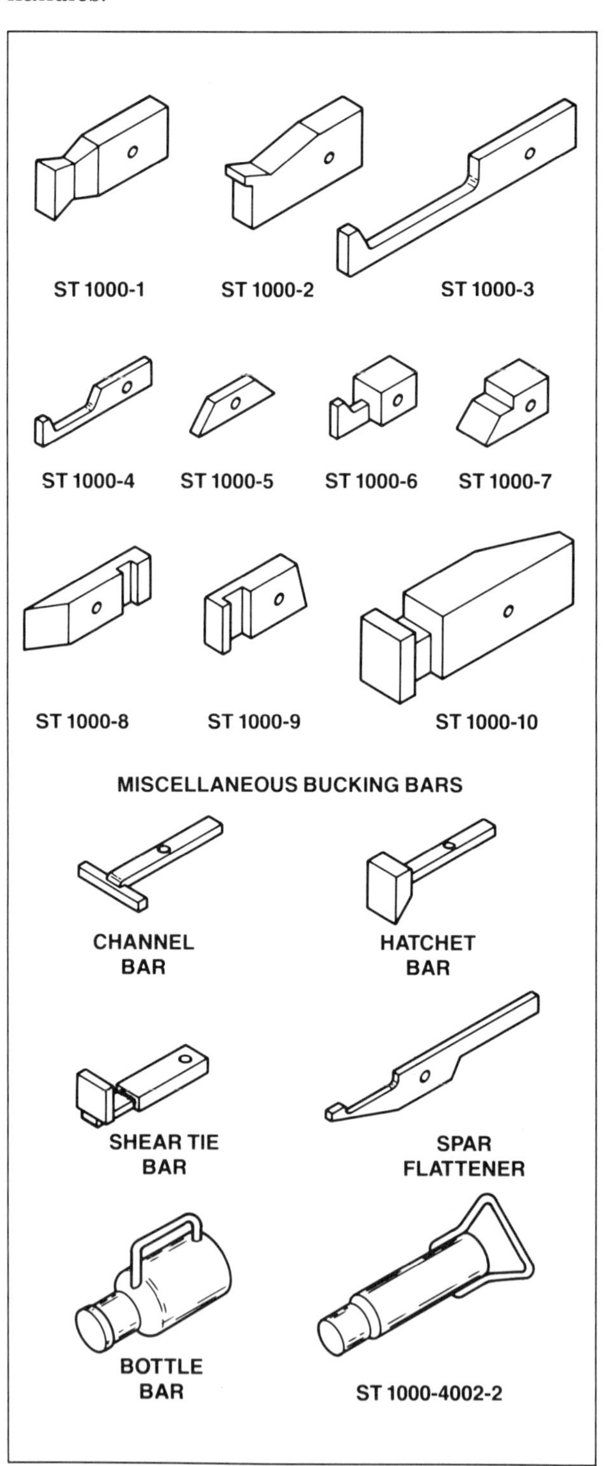

Fig. 2-13 Bucking bars.

RIVET — DRILL — CLECO — BUCKING BARS — MINIMUMS								
DIAMETER OF RIVET	RIVET DIAMETER DECIMAL	DRILL DECIMAL	OVERSIZE OF DRILL	DRILL NUMBER	CLECO COLOR	BUCKING BAR WEIGHT	MINIMUM RIVET PITCH	MINIMUM EDGE DIST.
3/32	.0937	.0980	.0043	40	SILVER	2-3 lbs.	9/32	6/32
1/8	.125	.1285	.0035	30	COPPER	3-4 lbs.	3/8	2/8
5/32	.15625	.1590	.0028	21	BLACK	3-4.5 lbs.	15/32	10/32
3/16	.1875	.1910	.0035	11	GOLD	4-5 lbs.	9/16	6/16
1/4	.250	.2570	.007	F	GREEN	5-6.5 lbs.	3/4	2/4

NOTE: ALCOA CLAIMS THAT A RIVET SHOULD FIT AS TIGHT AS POSSIBLE BEFORE DRIVING, ESPECIALLY THOSE RIVETS OF HARDER ALLOY. THE HOLE CLEARANCES LISTED ABOVE ARE THE RECOMMENDED SIZES USED IN THE FIELD.

Fig. 2-14 Rivet information chart.

10. Bucking Bars

Bucking bars are tools used to form shop heads on solid-shank rivets during installation. Bucking bars are available in many sizes, shapes and weights. An assortment of bucking bars is illustrated in Figure 2-13. When installing rivets, it is important to remember that the bucking bar weight should correspond to the diameter of the rivet being driven (Figure 2-14).

The face of the bucking bar is smooth and must be protected from any nicks or scars which can affect the proper shaping of the driven head. Damaged bucking bar faces can be filed and then smoothed using oil and ***crocus cloth.***

11. Hand Files

Files are classified according to shape, length, and cut (Figure 2-15). The shapes of files are flat, triangular, square, half round, and round. The length of a file is measured from the heel to the point. That is, the tang is not included. Files may be single or double cut. A single-cut file has one set of teeth angled 75° to its center line. A double-cut file has two sets of teeth, one angled 75° and the other 45° to the center line of the file.

There are several filing techniques. The most common is to remove rough edges and slivers from the finished part before it is installed. Cross filing is a method used for filing the edges of metal parts which must fit tightly together. Cross filing involves clamping the metal between two strips of wood and filing the edge of the metal down to a preset line. Draw filing is used when larger surfaces need to be smoothed and squared. It is done by drawing the file over the entire surface of the work.

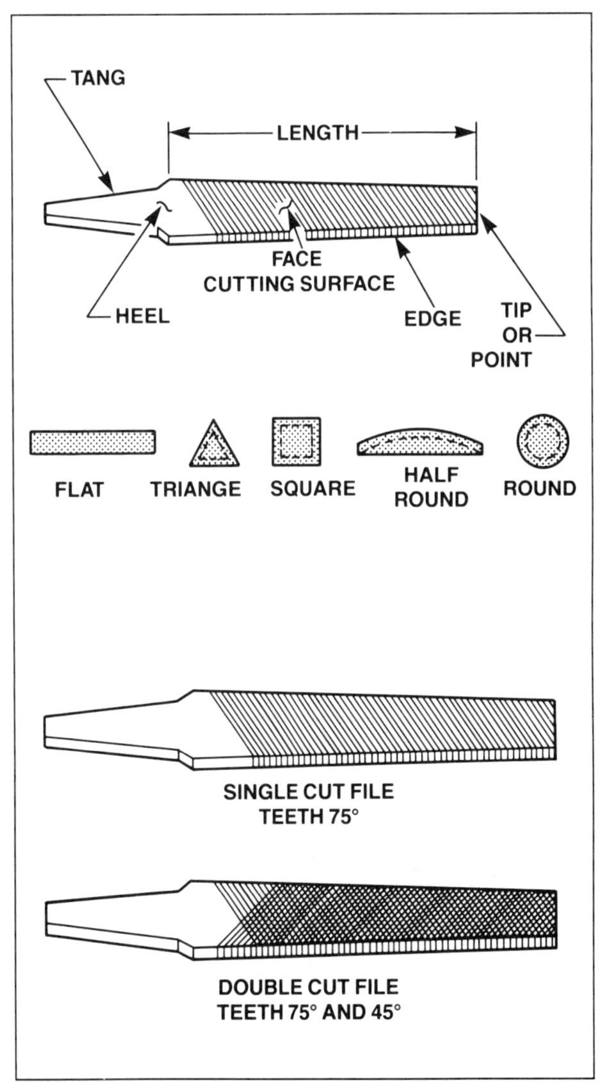

Fig. 2-15 File parts and shapes.

To protect the teeth, files should be stored separately in a plastic wrap or hung by their handles. Files which are kept in a toolbox should be wrapped in waxed paper to prevent rust from forming on the teeth.

File teeth can be cleaned with a file card. Attached to the side of the file card is a ***de-pinning*** pick, which is used to remove metal particles lodged in the gullets of the file teeth. Oil build-up can be removed from the gullets by rubbing chalk on the teeth. The oil will be absorbed by the chalk.

12. Hand Shears

Hand shears are needed to make large holes, curved parts, round patches, and ***doublers.*** They have colored handles which identify the direction of the cuts: Yellow cuts straight, green curves right, and red curves left (Figure 2-16).

13. Nibblers

Nibblers, as the name suggests, cut metal by nibbling away small pieces until the correct size is obtained. A typical hand nibbler (Figure 2-17) is operated by squeezing repeated notches out of the metal. Nibblers can be used to enlarge an inspection hole or to cut a notch to fit around an obstruction.

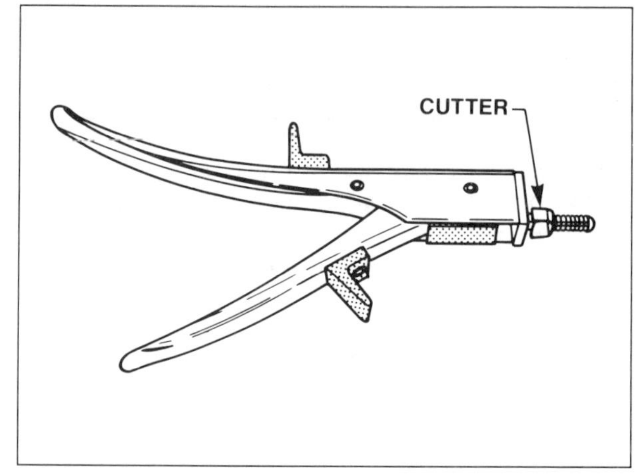

Fig. 2-17 Hand nibblers.

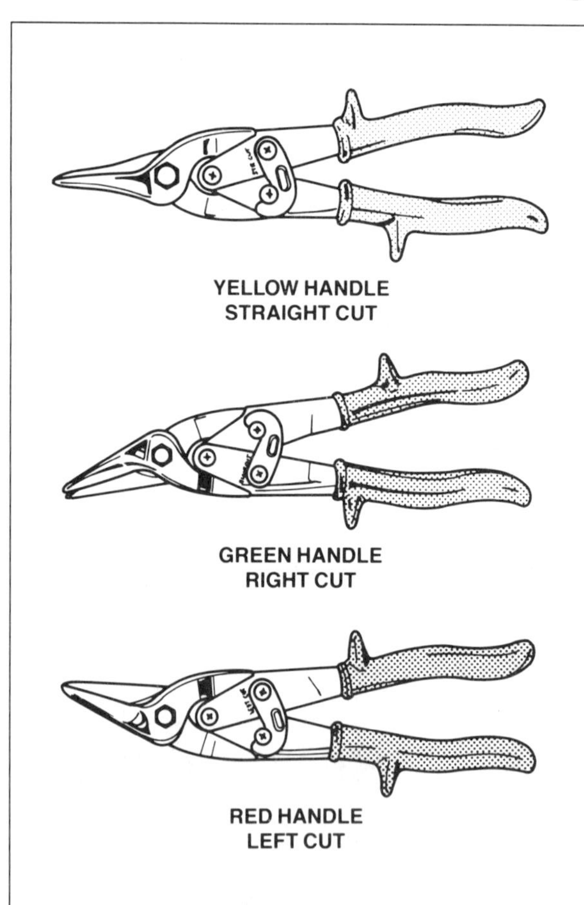

Fig. 2-16 Aviation snips.

14. Chassis Punches

The chassis punch, once used only for installing radios and other avionics appliances, has become a useful tool for the sheet metal mechanic because of its ability to make neat holes. Chassis punches are used to make lightening holes in newly formed ribs or access holes for inspection purposes.

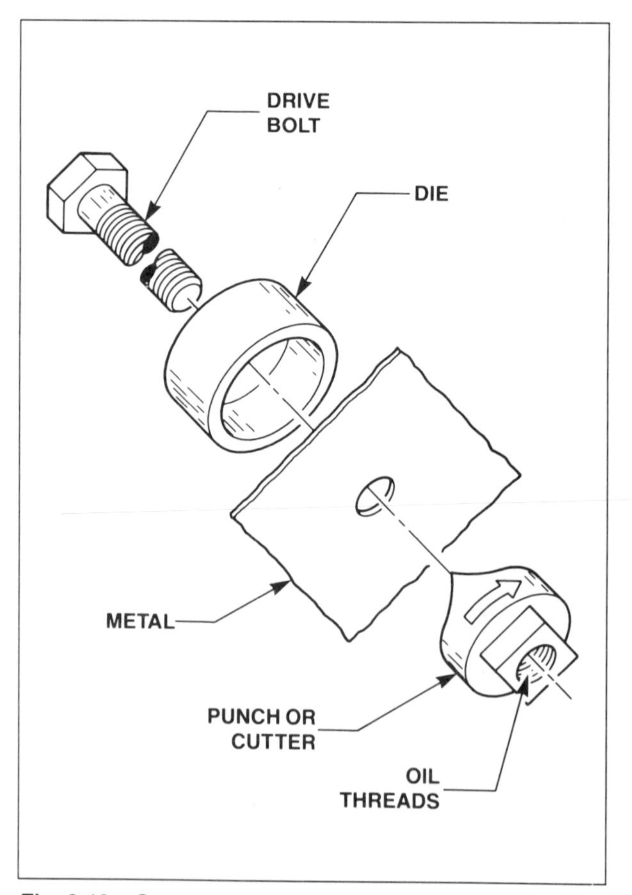

Fig. 2-18 Greenlee hole cutter.

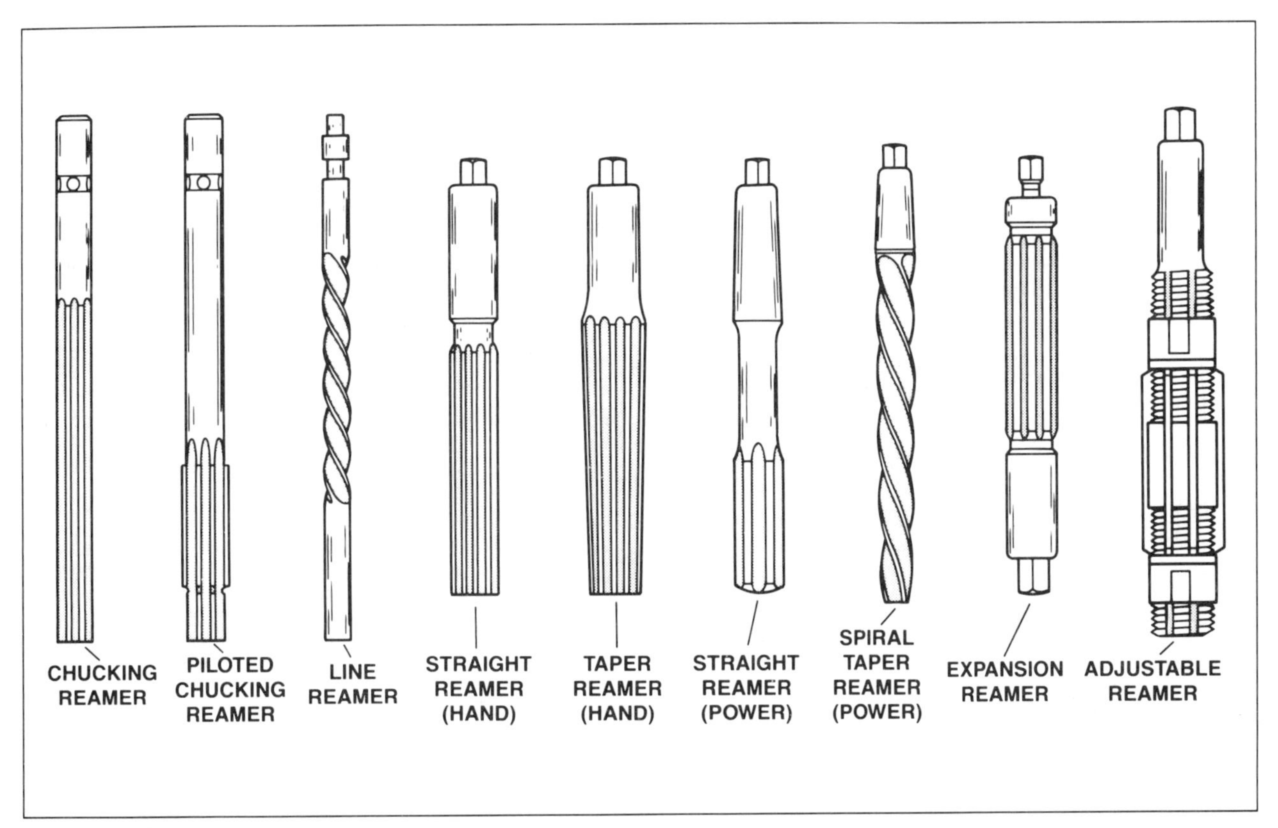

Fig. 2-19 Reamers.

The Greenlee chassis punch (Figure 2-18) can be used with a 3/8-inch drill and a 3/4-inch wrench to open holes from 1/2 to 1 1/4 inches in diameter. For larger holes, a larger drill and wrench will be needed.

15. Reamers

When a close tolerance fit is required for a special fastener, an undersized hole is first drilled, and the hole is then brought up to specifications by using a reamer. Reamers are driven by a drill motor or turned by hand. They are available in standard and adjustable sizes. Straight, tapered, spiral, expansion, and adjustable reamers are shown in Figure 2-19.

The correct way to use a reamer is to turn it only in the cutting direction when entering the hole or taking it out. If the reamer is turned in the opposite direction, the cutting edges will be damaged.

16. Routers

The router is not a tool normally used to repair aircraft structures. It is used primarily to remove damaged honeycomb skin and core material from the control surfaces of modern jet aircraft. Routers are driven either pneumatically or electrically. Each type uses the same straight cutting bits, either high-speed steel or carbide-tipped. The bit best suited for cleaning out damaged honeycomb is the carbide-tipped type.

Figure 2-20 shows the router with a cutting bit. The router bits range in width from 3/16 to 1 inch and from 3/4 to 1 1/2 inches in depth.

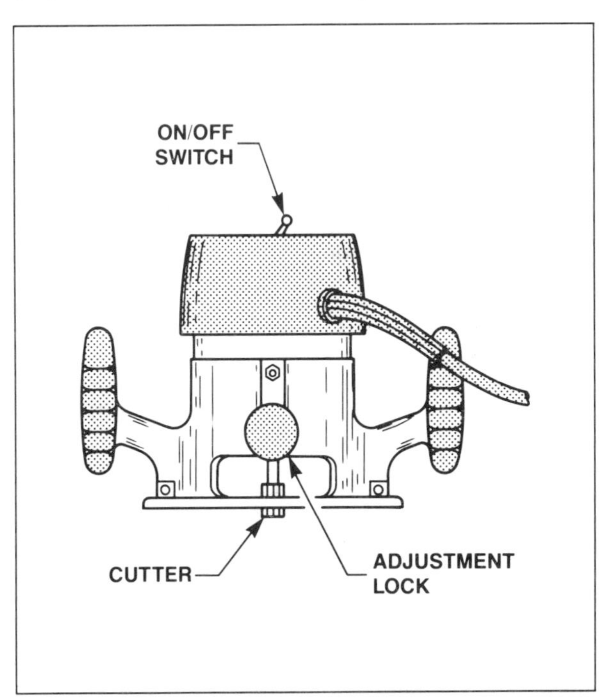

Fig. 2-20 Router.

17. Hole Finder Or Duplicator

Hole finders, or duplicators, are used to locate holes to be made in undrilled skin. Hole finders can be easily and inexpensively made to meet specific needs. Figure 2-21A shows a typical homemade hole finder.

The holes in the rib or stringer to be reused serve as the alignment holes. The duplicator marks the location of the holes to be made in the new skin. To use a hole duplicator, first insert the rivet alignment pin into one of the holes on the rib. Then place the undrilled skin over the hole finder's rivet head. Finally, center-punch the new skin (Figure 2-21B).

B. Bench And Floor Tools

1. Squaring Shears

Squaring shears can be operated either manually or hydraulically. A foot-operated treadle can be inexpensively converted to a hydraulic type by installing a hydraulic power pack on the machine. A hydraulic squaring shear can cut aluminum alloy up to 1/8 of an inch thick and aircraft steel up to 16 gauge.

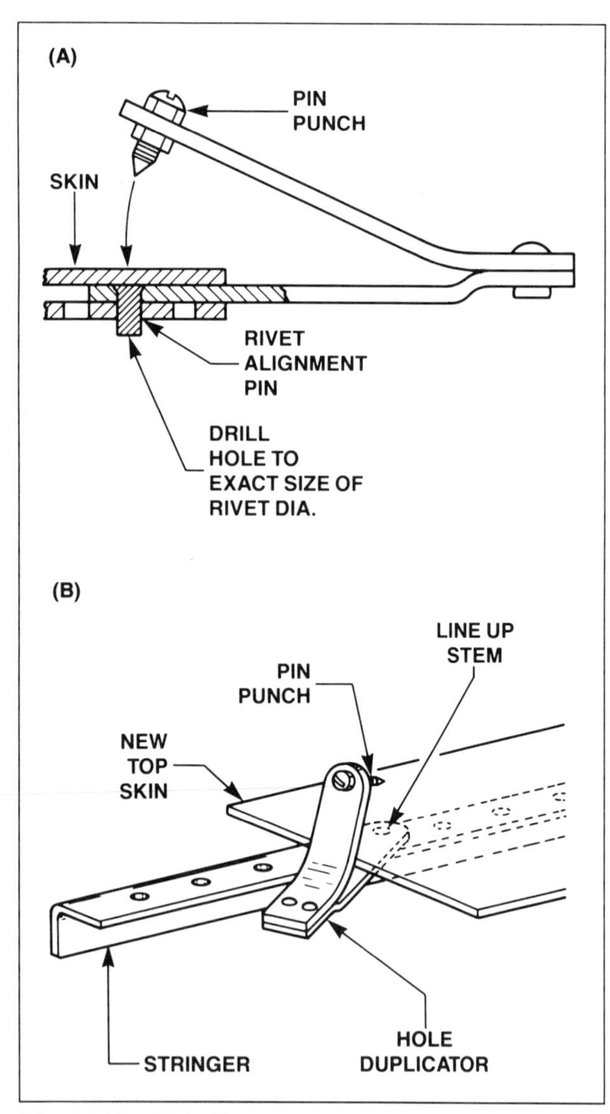

Fig. 2-21A Hole Finder.
Fig. 2-21B Hole finder used to duplicate holes.

Figure 2-22 shows a squaring shear. The primary purpose of the side fence is to square metal to the cutting blade. A piece of sheet metal can be squared by placing it beyond the cutter, shearing off one side, and turning the metal 90° to make another cut.

Extending from the back of the machine are two rods, which, like the side fence, have a measuring scale embossed on them. Connected to these two rods is a stop fence that can be adjusted for longer

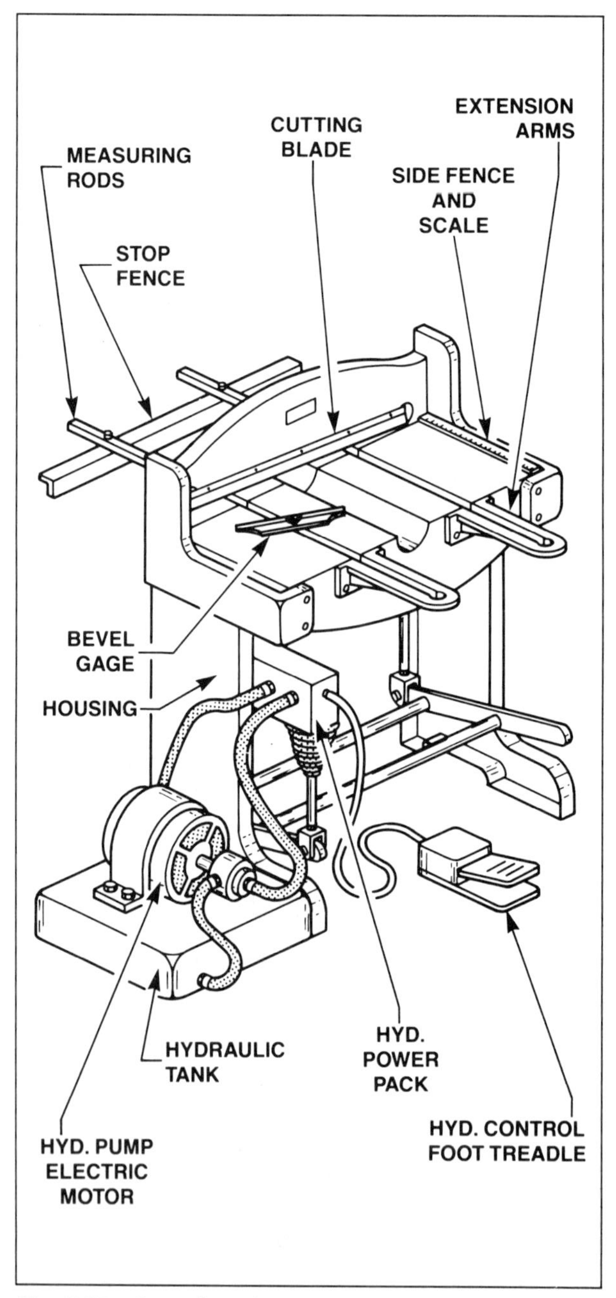

Fig. 2-22 Squaring shear.

cuts. The rods can be set to measure the distance from the cutting blade to the adjusted stop.

For example, if a sheet metal skin is to be cut 10 inches long and has to be squared, slide the skin along the fixed side fence, pushing it until it contacts the stop fence, which has been set at 10 inches. Then cut by stepping on the treadle.

2. Bending Brakes

There are many different sizes and types of bending brakes. The most versatile and widely used is the box and pan brake (Figure 2-23). The box and pan brake is fully adjustable and comes in various lengths. It has removable nose pieces that make the construction of a pan rib possible. The whole brake section can be moved backwards and forwards, allowing the addition of thicknesses of metal to the nose of the brake to increase the radius of the bend (Figure 2-24). As the ***bend radius*** is increased, the nose brake clearance must also be increased.

Straight bending brakes range from four to twelve feet in length. They are usually not adjustable, and it is difficult to install a bend radius bar on the nose of the brake. The straight bending brake is used to make long stringers.

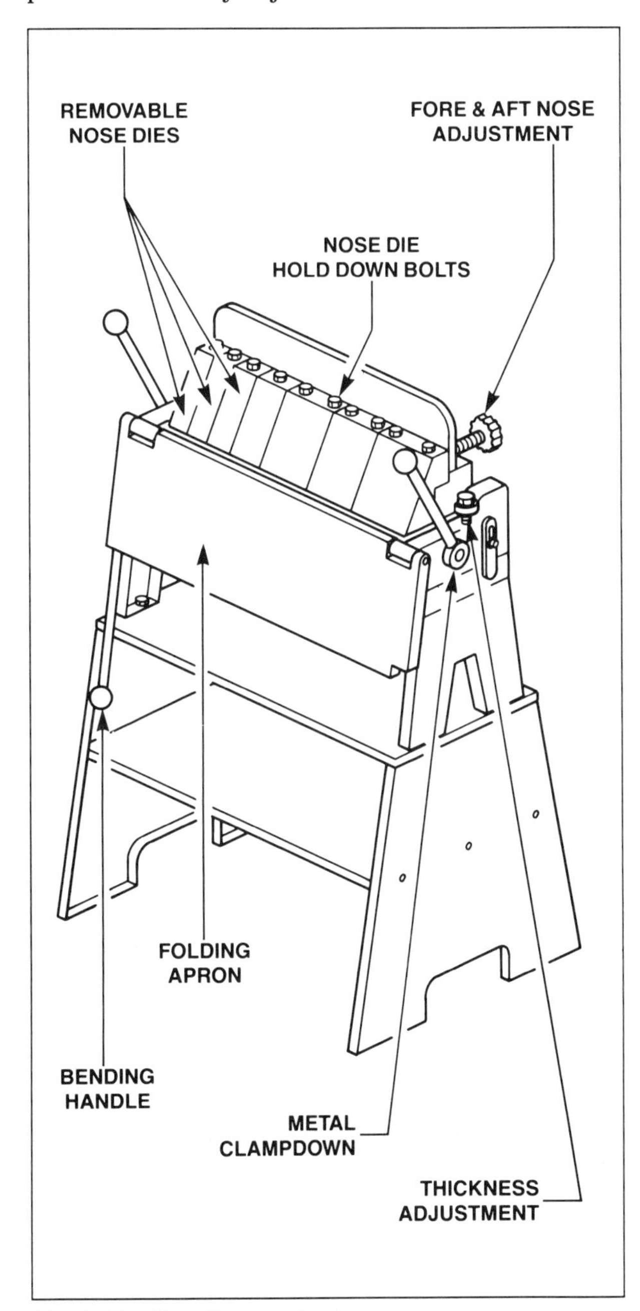

Fig. 2-23 Bending pan brake.

3. Slip Roller

A slip roller (Figure 2-25) is used to make a cylinder or a duct. The slip roller consists of three rollers interconnected by gears and turned by a hand crank. The rollers can be individually adjusted to produce a tight or a loose roll. After a metal cylinder is formed, it is removed from the roller by activating a release which allows the roller to slip free of its holder at one end. The part can then be removed.

4. Drill Press

A drill press is used to drill precision holes. Its use on skins is limited by the reach, or distance, between the center line of the spindle and the stand which holds the drill's drive mechanism. It is more commonly used to predrill holes in new spars or stringers.

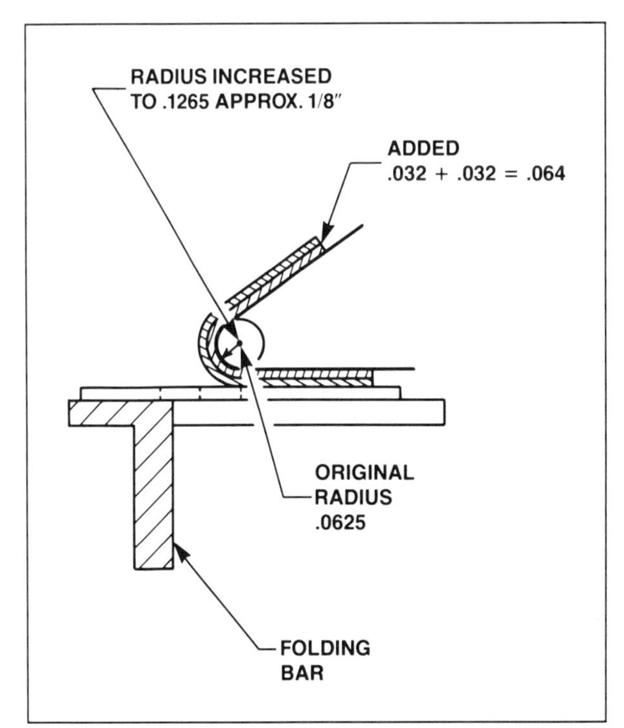

Fig. 2-24 Increasing bend radius bar.

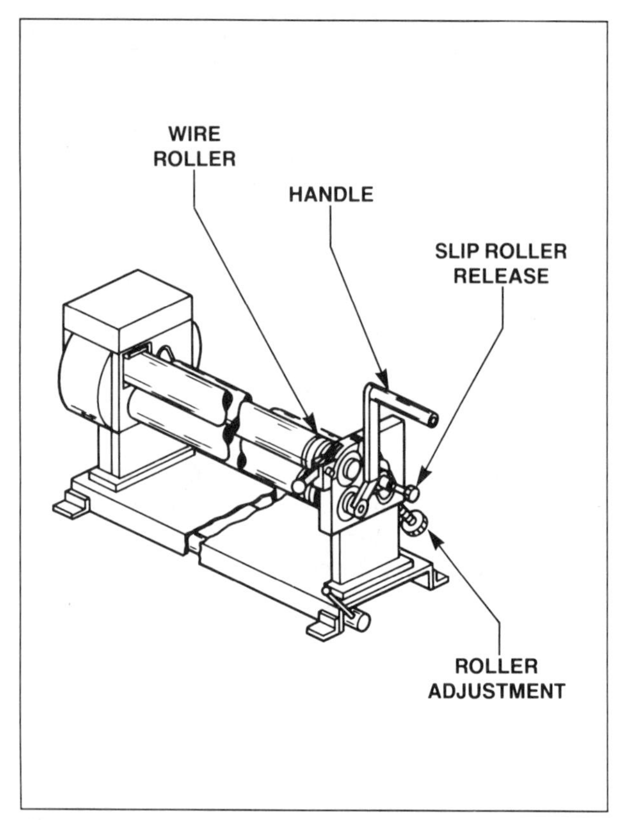

Fig. 2-25 Slip roll former.

The drill press has four main parts: Floor stand, adjustable table, adjustable spindle, and electric motor (Figure 2-26). The table can be raised upwards toward the spindle to reduce the distance that the spindle has to travel downward to meet the work. On some drill presses, the spindle has a threaded stop which can be adjusted to control the amount of downward travel and thus allow the drill press to be used as a stop countersink cutter. The drill press electric motor is bolted to the same frame as the spindle. Drill speed is determined by the belt's location on the motor and by the spindle pulleys. If low speed is required for drilling, the small pulley on the motor will be connected to the large pulley on the spindle.

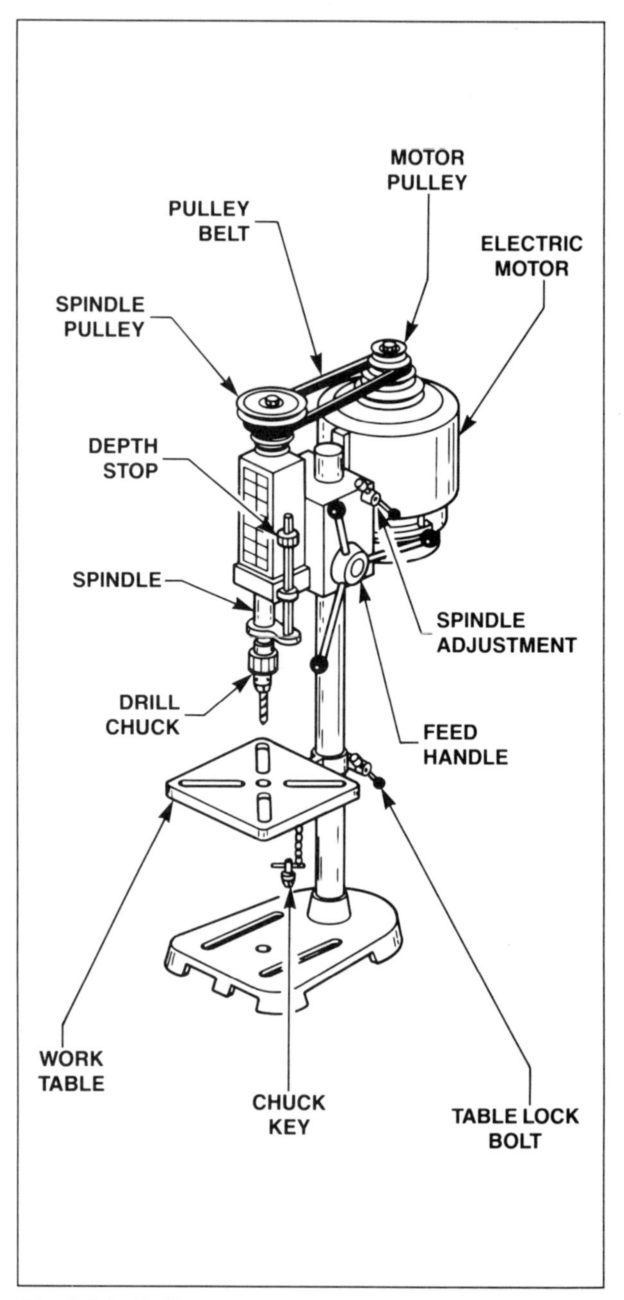

Fig. 2-26 Drill press.

5. Combination Band Saw

A combination hack and band saw (Figure 2-27) is used when thick sheet metal cannot be cut with the squaring shears. A medium-size combination saw will do most such cutting jobs.

6. Electric Grinders

The electric grinder is used regularly to sharpen dull or broken drills. A grinder does not have to be large; a small 1/2 to 3/4 horsepower motor is adequate. The grinding wheel best suited for sharpening drills is a fine grit stone. When a grinding wheel becomes grooved and uneven, it should be re-dressed and squared with a dressing tool. Care should be taken to avoid dropping the grinding wheel because it can develop small cracks which could later cause the wheel to fly apart and seriously injure the operator.

Note: Never use a grinding wheel that has been dropped or that has a flaw of any kind.

C. Airframe Special Tools

Tools used for aligning fuselages or supporting wings while they are being repaired are called

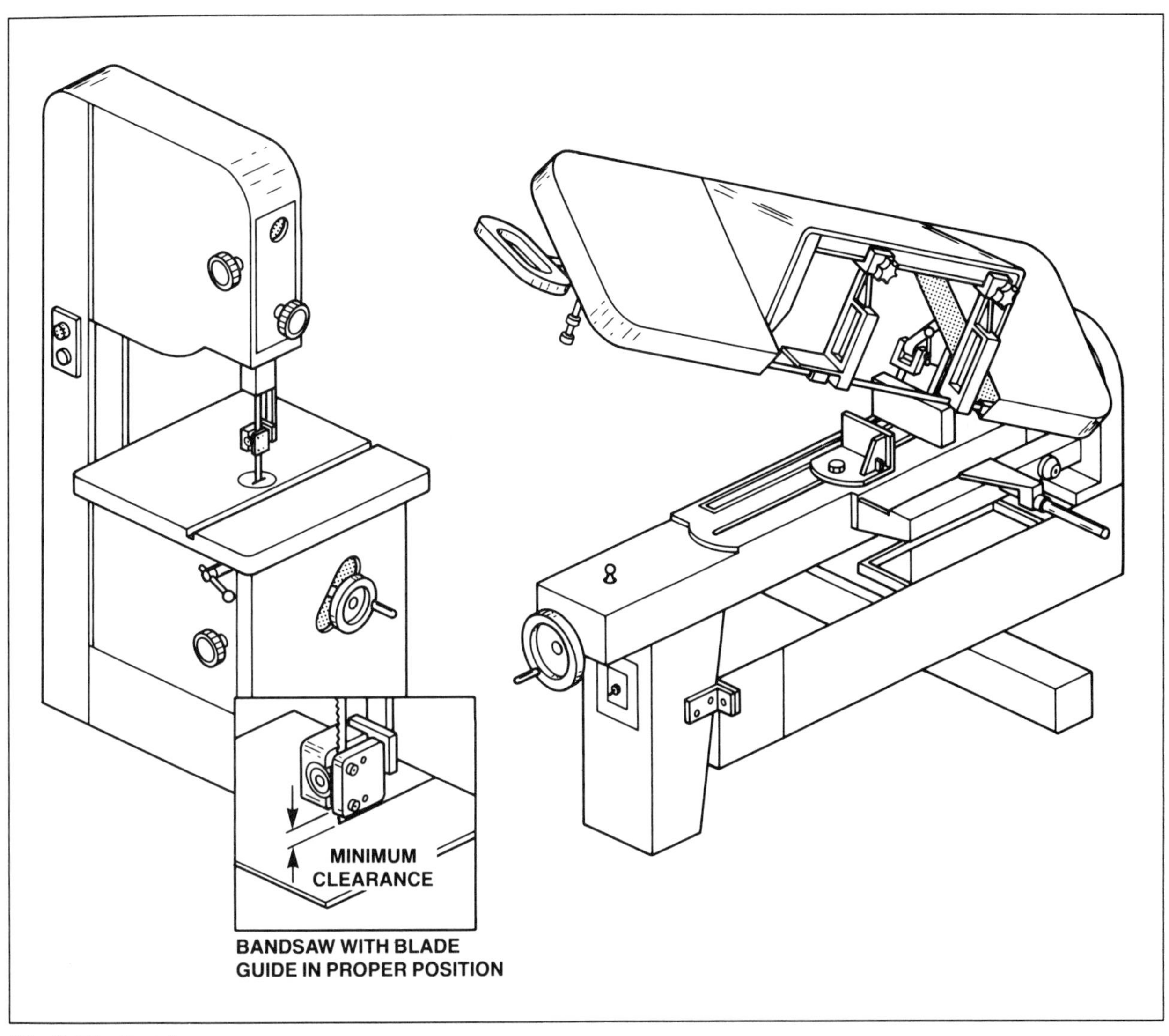

Fig. 2-27 Band saw.

slings or ***jigs.*** Most of the fuselage slings are shop-made for the specific repair to be done. For example, fuselage slings or saddles support the undamaged parts of the fuselage while the damaged parts are being removed. When a braced wing needs to be repaired and the strut must be removed, a wing jig or stand is placed underneath one of the compression ribs. Figure 2-28 shows examples of slings, jigs, and braces.

QUESTIONS:

1. *What is the included angle of a drill used to open holes on stainless steel?*
2. *What is the sharpening angle of a drill used to open holes on aircraft sheet metal?*
3. *What type of drill motor is the most safe to use around aircraft?*
4. *How many number drills are there in a normal set?*
5. *What size number drill is used to open holes for a 1/8-inch diameter rivet?*
6. *What are the three purposes of drill flutes?*
7. *What color Cleco clamp is used when installing 1/8-inch diameter rivets?*
8. *Name two types of gun sets used to install rivets on modern aircraft.*
9. *What weight bucking bar is needed to install a 5/32-inch diameter rivet?*
10. *Name the parts of a file.*
11. *What is a double-cut file?*
12. *How should reamers be removed from a hole?*
13. *What tool is used to remove damaged honeycomb skin?*
14. *How would you increase the radius of a pan bending brake?*
15. *If the belt were connected to the large pulley on the motor, would the drill press spindle speed be high or low?*

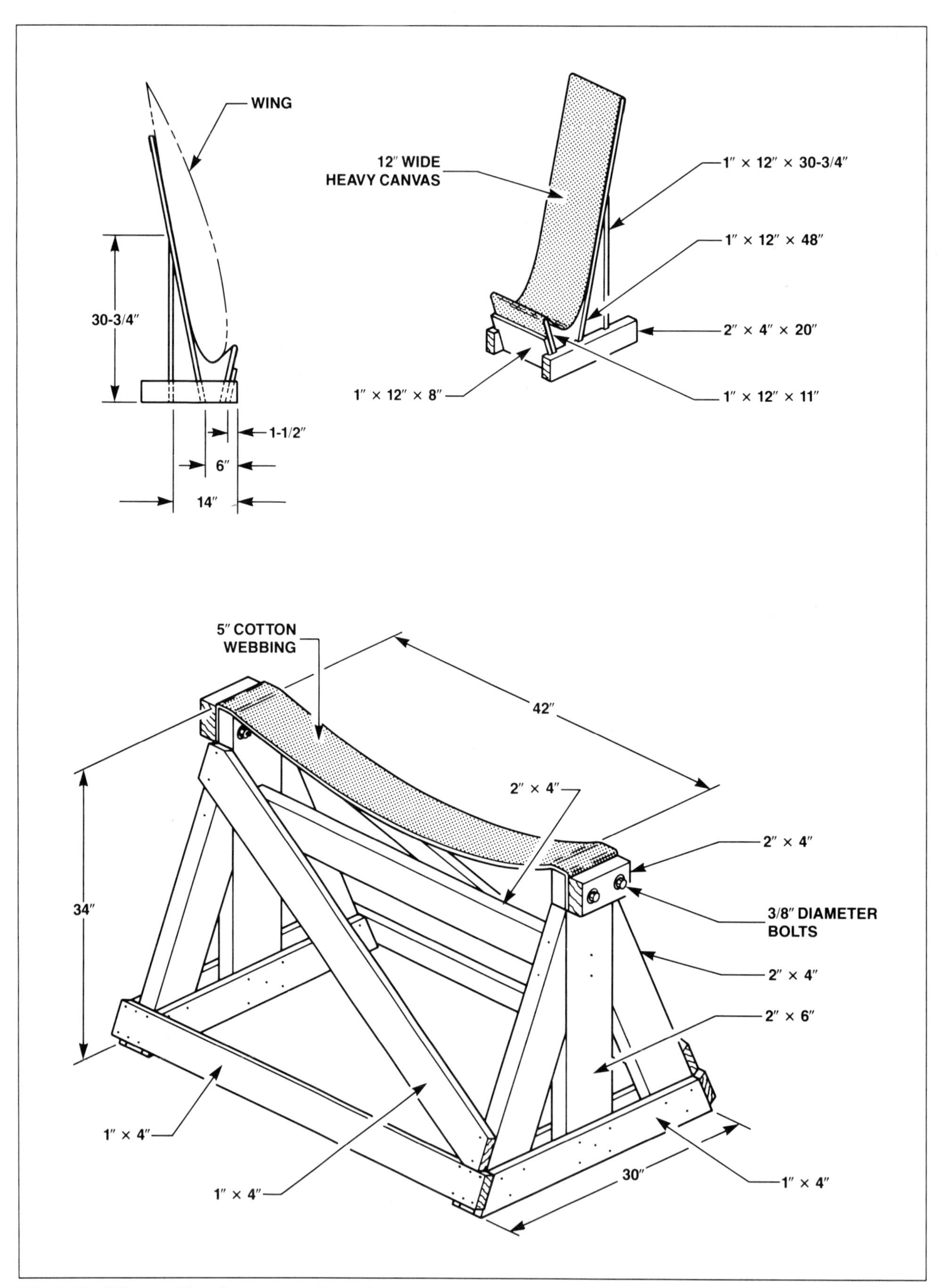

Fig. 2-28 Wing and fuselage support stands.

Chapter III
Aircraft Metals And Alloys

Modern aircraft are manufactured from many different types of metal alloys. Those most commonly used are aluminum alloy, titanium alloy, Monel (a nickel alloy), stainless steel, and chrome-molybdenum steel. Civilian aircraft are constructed primarily from heat-treated aluminum alloys, while military aircraft are constructed primarily from titanium and stainless steel.

A. Aluminum Alloys

Pure (99.0%) cast aluminum is unsuitable for aircraft structures because it is too soft. However, because of its light weight (one-third the weight of steel), it is, when alloyed with other metals or elements, an ideal structural material. Alloyed aluminum is produced in cast or wrought form. Cast aluminum has a grain structure that is very coarse; thus the metal is brittle. Cast aluminum is used in the construction of aircraft wheel castings and engine crankcases.

When aluminum is wrought, its grain structure is compressed and tightened as it is forced into shapes of plates, rods, extrusions, or skins. Wrought aluminum, used extensively in aircraft construction, is either non-heat-treated or heat-treated. Most structural aircraft parts are made of heat-treated aluminum alloys.

Alloying is mixing aluminum with other metals to make it stronger. The percentage of copper or zinc mixed with aluminum will determine the strength of the alloy. Stirring sugar into a cup of coffee is a good example of what happens when metals are alloyed. The degree of sweetness in the coffee is directly related to the amount of sugar added to the mixture. Similarly, the percentage of other elements in the alloy determines its characteristics. For example, when aluminum is mixed with copper, manganese, or magnesium, the alloy is capable of carrying major structural loads.

Figure 3-1 shows the percent of alloying elements mixed with aluminum to form the various types of metals used in the aircraft industry. The alloy is identified by the predominant alloying ingredient. For example, the alloy 2024, which contains 4.5% copper, is a copper-bearing alloy

ALLOY	PERCENT OF ALLOYING ELEMENTS — ALUMINIUM AND NORMAL IMPURITIES CONSTITUTE REMAINDER								
	COPPER	SILICON	MANGANESE	MAGNESIUM	ZINC	NICKEL	CHROMIUM	LEAD	BISMUTH
1100	—	—	—	—	—	—	—	—	—
3003	—	—	1.2	—	—	—	—	—	—
2011	5.5	—	—	—	—	—	—	0.5	0.5
2014	4.4	0.8	0.8	0.4	—	—	—	—	—
2017	4.0	—	0.5	0.5	—	—	—	—	—
2117	2.5	—	—	0.3	—	—	—	—	—
2018	4.0	—	—	0.5	—	2.0	—	—	—
2024	4.5	—	0.6	1.5	—	—	—	—	—
2025	4.5	0.8	0.8	—	—	—	—	—	—
4032	0.9	12.5	—	1.0	—	0.9	—	—	—
6151	—	1.0	—	0.6	—	—	0.25	—	—
5052	—	—	—	2.5	—	—	0.25	—	—
6053	—	0.7	—	1.3	—	—	0.25	—	—
6061	0.25	0.6	—	1.0	—	—	0.25	—	—
7075	1.6	—	—	2.5	5.6	—	0.3	—	—

Fig. 3-1 Nominal composition of wrought aluminum alloy.

of aluminum. The 7075 alloy, which contains 5.6% zinc, is a zinc-bearing alloy of aluminum. Figure 3-2 shows how to read the alloy codes.

1. Copper (Cu) 2XXX

When copper is mixed with aluminum, the aluminum becomes more malleable and ductile. ***Ductility*** is the ability of the metal to be drawn into wire and bar stock. Pure aluminum is also malleable and ductile, but the addition of copper enhances these physical properties. The copper also acts to prevent stress cracks from forming while the metal is worked and makes some alloys, like 2024T3, shock resistant.

Copper, which melts at 1981°F, is one of the most ductile of all the metals. It can, however, be made harder by ***coldworking.***

2. Manganese (Mn) 3XXX

Manganese is a gray-white, brittle metal which melts at 2273°F. When manganese is mixed with aluminum, it provides a surface highly resistant to wear and ***corrosion.***

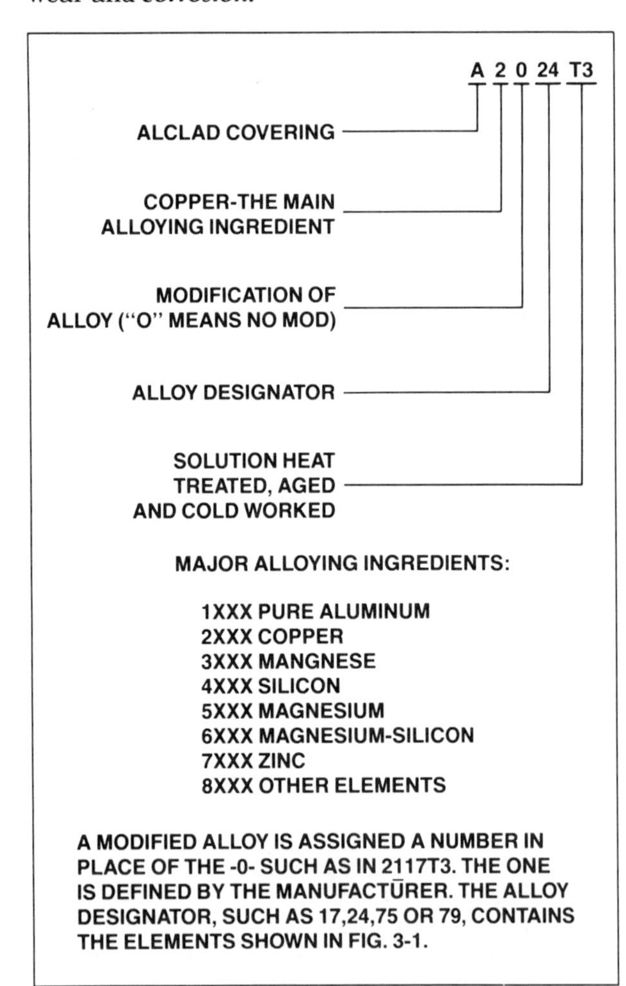

Fig. 3-2 Aluminum alloy identification.

3. Silicon (Si) 4XXX

Although silicon, which melts at 2538°F, is not metallic, it has properties which, in alloy, make aluminum harder but not brittle.

4. Magnesium (Mg) 5XXX

Magnesium weighs 2/3 as much as aluminum. It is strong enough to use structurally only when it is alloyed with aluminum, zinc, or manganese. Also, because magnesium is very corrosive and burns easily — especially in ribbon or powder form — it is seldom used in sheet form. However, the aluminum-magnesium 5056 rivet is commonly used to hold the skins onto magnesium control surfaces.

5. Zinc (Zn) 7XXX

When zinc is mixed with aluminum, the resulting alloy is stiffer and more brittle than pure aluminum. Zinc — a bluish-white, lustrous metal — is brittle at room temperature but malleable when heated. Zinc melts at 786°F.

B. Wrought Aluminum Alloys

Wrought aluminum is divided into two categories: Non-heat-treatable and heat-treatable. After the alloy is wrought, it is in an "as fabricated" (F) condition until it is heat treated or annealed. For example, the alloy 2024 would be 2024F before being heat treated.

1. Non-Heat-Treatable Aluminum

Non-heat-treatable aluminum alloys are hardened by alloying and ***strain hardening.*** The terms "coldworked", ***"work hardened",*** and "strain hardened" refer to hardening processes such as rolling, drawing, bending, or pressing. These alloys cannot be hardened by heat treatment.

Non-heat-treated aluminum alloys that are work hardened are designated by the letter H and the number 1, 2, or 3. H1 is strain hardened only; H2 is strain hardened and partially annealed; and H3 is strain hardened and then stabilized. Following these designations is one of four numbers — 2, 4, 6, or 8 — which indicates the degree of hardness of the alloy. For example, the alloy 3003H28 is strain hardened and stabilized to its fullest degree of hardness.

The four most commonly used non-heat-treatable aluminum alloys are 1100, 3003, 5052, and 5056.

The 1100 alloy is 99.0% pure. It is not used to make structural parts for aircraft. It is used to make small-diameter, low-pressure tubing, rivets, and reciprocating engine baffles.

The 3003 alloy has as its major alloying ingredient manganese, which gives it moderate

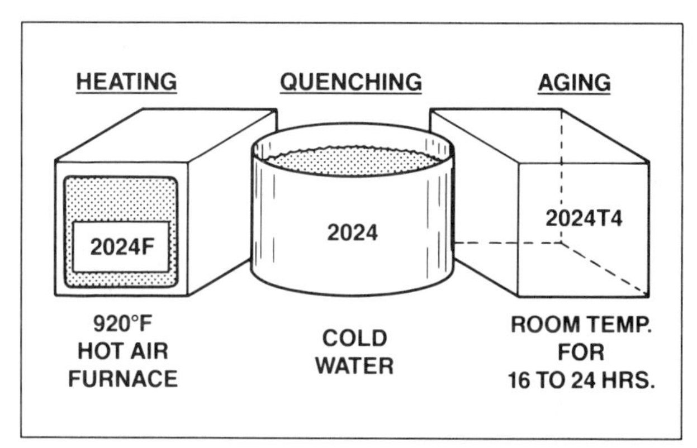

Fig. 3-3 Heat treatment with natural age hardening. Heating time varies according to thickness of metal. Time period between heating and quenching must be accomplished as soon as possible to reduce the possibility of causing intergranular corrosion.

strength and excellent ***formability.*** This alloy is used to construct wing tips and prop spinners.

The 5052 alloy has as its major alloying ingredient magnesium, but it also contains a slight amount of chrome. This alloy has excellent structural strength and good formability. It maintains its properties well at elevated temperatures. It is used to make low-pressure tubing and storage tanks for hydraulic fluids, fuel, oil, and other liquids.

The 5056 alloy also has magnesium as its major alloying ingredient. It is primarily used as rivet stock for magnesium control surface skins. Rivets made from 5056 are magnesium compatible and therefore will not cause corrosion when installed into magnesium skins.

2. Heat Treatment: Natural Aging

The heat-treated aluminum alloys are extensively used in aircraft structures. Commonly used heat-treatable alloys naturally age hardened are 2117, 2017 and 2024.

There are three stages in the heat treatment process: Heating, quenching, and aging. In its "as fabricated" condition, the aluminum is placed in a heat-treating oven and heated for a period of time determined by its alloying elements and by its size (Figure 3-3).

For example, the aluminum alloy 2024F is heated to a temperature of 920°F and held at that temperature until all parts are equally heated. After being heated, the metal is quickly transferred to a quenching tank. The quenching operation must be performed as quickly as possible in order to reduce the possibility of intergranular corrosion. After quenching, the aluminum alloy is kept at room temperature for 16 to 24 hours to ensure that the metal has age hardened. Age hardening, the final step in the heat-treatment process, makes the metal naturally hard.

The designation of metal so treated changes from F, as fabricated, to T4, heat treated and age hardened. If the metal is further hardened by coldworking (that is, mechanically) its designation is T3. See Figure 3-4 for other temper designations of heat-treatable and non-heat-treatable aluminum alloys.

HEAT TREATING THE COMMON ALUMINUM ALLOYS

	SOLUTION HEAT TREATMENT		PRECIPITATION HEAT TREATMENT			ANNEALING TREATMENT	
ALLOY	METAL TEMP.	TEMPER DESIGN.	METAL TEMP.	APPROX. TIME OF HEATING	TEMPER DESIGN.	METAL TEMP.	HEATING TIME IN HOURS
1100-O	—	—	—	—	—	650°F	—
3003-O	—	—	—	—	—	775°F	—
5052-O	—	—	—	—	—	650°F	—
5056H32	—	—	—	—	—	—	—
2117⋆	940°F	T4	—	—	—	775°F	2-3
2017	940°F	T4	—	—	—	775°F	2-3
2024	920°F	T4	375°F	7-9	T86	775°F	2-3
6061	970°F	T4	350°F	6-10	T6	775°F	2-3
7075	870°F	W	250°F	24-28	T6	775°F	2-3
7050T73⋆	890°F#	W	250°F*	4 HRS.	T73	355°F*	8 HRS.

NOTE 1: ⋆ means: alloy is for rivet stock only; #—890°F plus or minus 10°F; and *—plus or minus 5°F.

NOTE 2: All of the non heat treated alloys listed above are in the annealed form and the letter "O" is designated after their alloy codes, eg. 1100-O, 3003-O, and 5052-O.

Fig. 3-4 Temper designations of heat-treatable and non-heat treatable aluminum alloys.

Heat-treatable aluminum alloys commonly used in the construction of modern aircraft are 2117T4, 2017T4 and 2024T4 or T3.

The alloy 2117T4 is manufactured only as rivet stock. It is the most widely used rivet alloy in the aviation industry. The 2117T4 rivet is driven in the condition received from the manufacturer without any further treatment.

The 2017T4 alloy is manufactured in two similar forms. The first form dates back to 1925, when it was called 17ST and was used for rivets and structural skins. When being driven in diameters larger than 3/16 of an inch, the rivet stock frequently cracked. Also, it had to be re-heat treated and driven in a softened condition.

Recently, Alcoa reworked 2017T4 by slightly reducing the amount of magnesium and adding small amounts of iron and silicon. The new crack-free 2017T4 meets the original 2017 standards and can be driven in the condition received. It is used in the construction of the new Boeing 757 and 737-200 series aircraft.

The alloy 2024T3 is widely used for skin covering and internal parts of all types of aircraft. It has exceptional resistance to fatigue loads, it is highly resistant to cracks, it can withstand heavy load limits, and it retains high strength after damage. As rivet stock, however, the 2024T4 rivet is being challenged by a new rivet, the ***(Boeing rivet)*** 7050T73.

3. Precipitation Heat Treatment: Artificial Aging

Aluminum alloys containing zinc, magnesium, silicon, or copper are given a ***precipitation heat treatment*** after natural heat treatment is completed. For example, the alloy 7075 is given a normal heat treatment at 870°F and quenched in cold water. After it is precipitation heat treated at 250°F for 24 hours, it becomes 7075T6. Alloys are precipitation heat treated by heating them in an oven; time and temperatures vary (Figure 3-4).

This treatment has the effect of locking together particles in the grain of the metal, thus increasing strength, stability, and resistance to corrosion. Natural heat treatment begins the grain-binding process; precipitation heat treatment completes it. In addition, artificially aged alloys are generally over-aged to increase their resistance to corrosion, especially if, like 2024, they are subject to intergranular corrosion.

Metals which are given precipitation heat treatment usually lose some malleability and ductility, and their mechanical properties are so changed as to reduce their ability to be reshaped cold without cracking.

The most commonly used precipitation heat-treated alloys are those containing zinc. The alloy 7075T6 has high impact resistance and therefore is used where great strength is required. The 7079T6 aluminum alloys are excellent for making forgings for heavy channels that carry landing gears or flaps of large aircraft. The alloy 7178 is used where compression loads are the greatest, for example in the ***superstructure*** of wide-body jets.

The alloy 7050T73 is the newest aluminum alloy. It was developed in 1979 by Alcoa and the US Air Force. A combination of aluminum, zinc, and magnesium, it is primarily used as a solid-shank rivet. The 7050T73 alloy is the strongest of any rivet alloy in use today. It has a high resistance to stress corrosion and is much stronger than the alloy 2024T31, which it has replaced on some modern jetliners.

Figure 3-4 lists heat-treatable and non-heat-treatable aluminum alloys and their temper conditions.

C. Titanium Alloys

Because titanium is light (it weighs half as much as steel) and strong, particularly when alloyed, it is much in demand for structural parts on high-speed aircraft. Because it remains strong up to 800°F, it is better than aluminum for use around hot sections of jet aircraft.

Titanium resists fatigue, cracking, fracturing, and corrosion. Alloyed, it does not need to be coated to prevent corrosion, but, because it is cathodic to magnesium and aluminum, it should be insulated from them.

One form of titanium alloy that is used in the manufacture of special fasteners is 6AL-4V: 6AL means that 6% of the alloy is aluminum; 4V means that 4% is vanadium. This alloy has a melting temperature higher than that of steel, and it is 56% more dense. It has a ***tensile strength*** equal to that of steel and twice that of aluminum — 180,000 PSI in the heat treated condition.

D. Monel

Monel is a nickel alloy which has the properties of high strength and excellent corrosion resistance. Monel contains 68% nickel, 29% copper, 0.2% iron, 1% manganese, and 1.8 % other materials.

Monel is used for the construction of sprockets and chains for landing gears, in the manufacturing of certain aircraft fasteners, and, in general, wherever both strength and high resistance to corrosion are needed.

E. Stainless Steel: 18-8 Steel

Stainless steel is used in many places on modern aircraft, such as ***fire walls,*** skins, structural parts, and special fasteners. Stainless steel is often called 18-8 steel, because it contains 18% chrome alloyed to 8% nickel and 74% steel. Chrome forms an oxide which, after buildup, prevents corrosion from taking place.

Stainless steel is strongest when it is cold rolled. It resists drilling by conventional means. When drilling stainless steel, the drill must be sharpened to an included angle of 140° and the tip speed must be slow, 750 RPM. Use steady drill pressure.

F. Chrome-Molybdenum Steel

Chrome-molybdenum (4130) steel, also called "chromally", is made up from plain steel, molybdenum, and chrome. The main alloying ingredient in chrome-molybdenum steel is chrome. In the code 4130, the first digit (4) indicates an amount of molybdenum; the second digit (1) indicates the amount of chrome; and the last two digits (30) specify, in hundredths of a percent, the amount of carbon in the alloy. There are three classes of carbon steels: low (.05 to .25%), mild (.25 to .65%), and high (.65 to above 1.0%). Because 4130 is a mild-carbon steel, it is easy to weld.

The alloy 4130 steel is most commonly used to construct engine mounts and shock struts. Chrome-molybdenum is highly resistant to shock and corrosion.

QUESTIONS:

1. *Which type of metal is most commonly used in the construction of modern civilian aircraft?*
2. *What is the major alloying ingredient in 7050T73?*
3. *What percentage of copper does the alloy 2024T4 contain?*
4. *What percentage of copper does the alloy 1100 contain?*
5. *What characteristic of zinc makes it a good alloying ingredient?*
6. *List the following metals in order of increasing weight: magnesium, iron, titanium, stainless steel, aluminum.*
7. *What is the major alloying ingredient in the aluminum alloy 2024T4?*
8. *What is the major alloying ingredient in the aluminum alloy 5056H32?*
9. *How can the strength of the alloy 5052 be increased?*
10. *How can the strength of the alloy 2024T4 be increased naturally?*
11. *What are the three steps necessary to heat treat and naturally age harden aluminum alloy?*
12. *What does the temper code T3 mean when it follows an alloy designation like 2117?*
13. *What is the precipitation heat treatment temperature for the alloy 7050T73?*
14. *What does the letter "O" following the alloy designation 3003-O mean?*
15. *What alloys are usually precipitation heat treated?*
16. *What does the code A2024T3 designate?*
17. *List three advantages of titanium alloy.*
18. *What are three major uses of the alloy 18-8 steel, and what is its trade name?*
19. *What does the code 6AL-4V designate?*
20. *What is the code number for chrome-molybdenum?*

Chapter IV
Aircraft Solid-Shank Rivets

Only the solid-shank rivet increases, when it is properly installed, in size and strength. A steel bolt, for example, will actually decrease in diameter when it is installed and torqued. Some manufacturers of special fasteners attempt to duplicate the natural action of solid-shank rivets by putting expanding sections on the fasteners' pulling stems. Most of these special fasteners only approximate the natural shank expansion of a driven solid-shank rivet.

Since pure aluminum weighs one-third as much as steel, aluminum alloy solid-shank rivets are lighter than many other fasteners. Their lightness is an advantage, but it limits their usefulness: Solid-shank rivets greater than 1/2-inch in diameter are not used. However, the allowable range, 3/32 to 1/2-inch in diameter, is broad enough for the needs of most typical aircraft construction and repairs.

A. Riveting

Riveting is sometimes referred to as a "mushrooming" process. When the rivet is being driven, certain physical changes take place. The rivet diameter cross-sectional area increases, the hardness of the rivet increases due to coldworking, the manufactured head expands, and the shank expands to the size of the hole. The result of this mushrooming process is that the rivet is coldworked, changing its temper designation from T4 to T3.

There are three steps involved in planning and executing a solid-shank rivet joint for structural repair: Layout, installation, and inspection. Figure 4-1 is a chart showing the rivet sizes, drill sizes, Cleco colors, bucking bar weights, minimum ***edge distances,*** and minimum ***rivet pitches*** for layout and installation of the commonly used solid-shank rivets.

Rivet spacing is important when laying out rivets to obtain a joint which is structurally sound and aesthetically balanced. The layout of rivets may be in rows abreast or transverse. One of the advantages of placing rivets in ***transverse rows*** is that it reduces rivet failure along the metal's grain structure. Rivets laid out in rows abreast have a greater tendency to fail along the grain.

Rivet pitch is the distance between one rivet or row of rivets and the next rivet or row of rivets. (Some handbooks refer to rivet pitch as gage.) Minimum rivet pitch is 3D, or 3 diameters, of the rivet being driven. (The diameter of a rivet is expressed as D and the length as L.) Average rivet pitch is 6 to 8 diameters. Maximum rivet pitch is 24 times the thickness of the top sheet of metal. Minimum transverse rivet pitch is 2.5 diameters of the rivet being driven.

Edge distance is the distance from the edge of the metal to the center of the first rivet or row of rivets. On aircraft, the minimum edge distance is 2D and the maximum is 4D. If an edge distance is larger than 4D, the edges may curl upwards and not lie flat and binding. When aircraft rivets are installed using less than 2D edge distance, the bearing edge strength of the metal will weaken.

Figure 4-2 illustrates edge distance and rivet pitch.

After the layout is computed, the first installation operations are to center punch the holes,

RIVET DIAMETER	DRILL SIZE	CLECO COLOR	BUCKING BAR WEIGHT	EDGE DISTANCE	RIVET PITCH
3/32	40	SILVER	2-3.0 LBS	6/32	9/32
1/8	30	COPPER	3-4.0 LBS	1/4	3/8
5/32	21	BLACK	3-4.5 LBS	10/32	15/32
3/16	11	GOLD	4-5.0 LBS	6/16	9/16

Fig. 4-1 Layout and installation of solid-shank rivets.

to select the correct drill, and to drill the holes to the proper size. All rivet holes must be center punched in order to prevent the drill from walking over the surface of the metal and defacing it. The indentation made by a center punch must be hard enough to catch the point of the drill, yet light enough to prevent denting the surrounding metal. Proper drill selection depends upon the size of the rivet being used. The hole for a solid-shank rivet is drilled approximately .002 to .004 of an inch larger than the nominal rivet diameter.

A rivet that is driven into a properly prepared hole needs to be sized according to diameter and length so that a correct size ***bucktail*** can be formed. Figure 4-3 shows the width and height of a normally driven bucktail.

The use of thin skins on many light aircraft requires that the upset rivet head be .66 times the diameter of the rivet high, and 1.33 inches times the diameter of the rivet wide. To determine the rivet length for a particular job, the thickness of all the metal parts must be known. All the individual thicknesses of the metal are referred to as the ***grip length.*** The grip length plus 1.5D is the proper length of the rivet (Figure 4-4).

1. Rivet Guns

The hand tool used to drive a rivet is called a pneumatic rivet gun or rivet hammer. Rivet guns are normally powered by compressed air and are classified as light-, medium-, or heavy-hitting. A light-hitting gun is used to install 3/32 and 1/8-inch diameter rivets. Medium-hitting guns are used

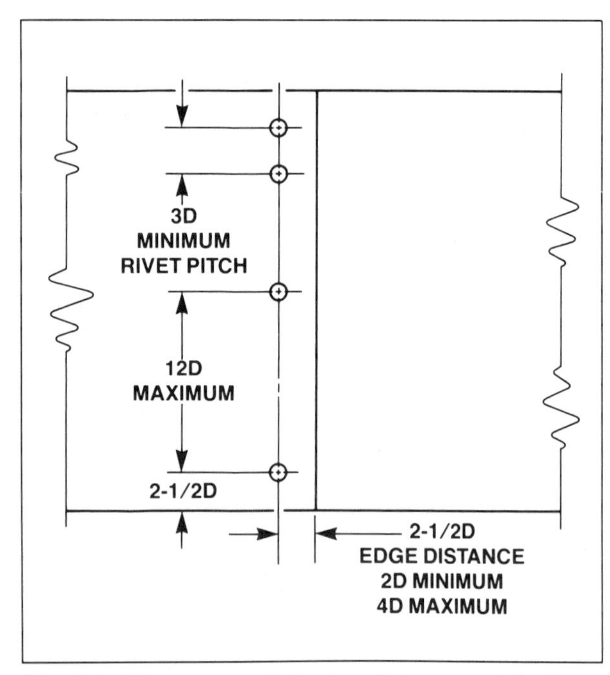

Fig. 4-2 Measurement of edge distance and rivet pitch.

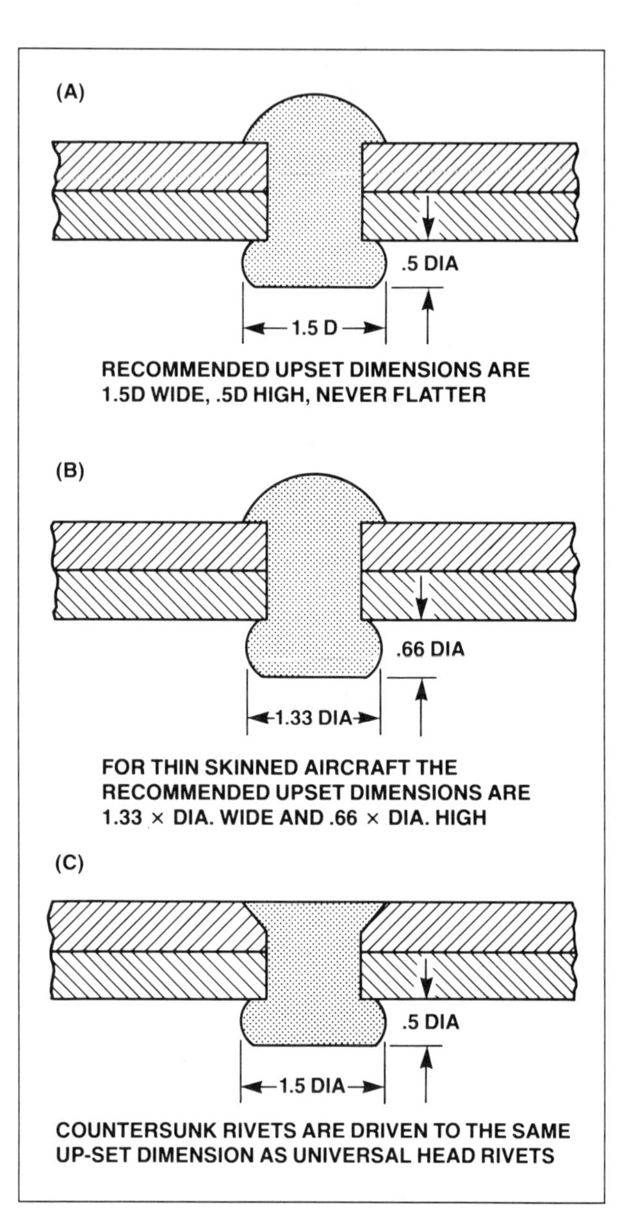

Fig. 4-3 Bucktail dimensions.

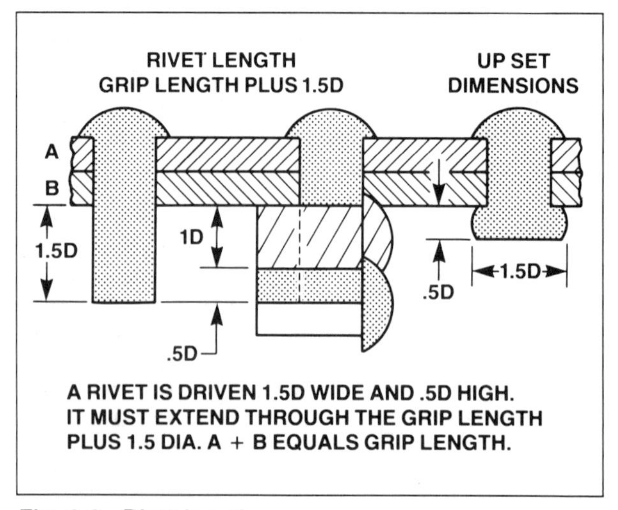

Fig. 4-4 Rivet length.

to install 5/32 and 3/16-inch diameter rivets. Heavy-hitting guns are used to install larger diameter rivets and some special fasteners.

There are two types of gun sets, one for ***universal head rivets*** and one for countersunk. The universal gun set is sized to fit the various shapes of manufactured heads on the rivet's driven end. The opposite end of the universal gun set fits into the rivet gun barrel and is held in place by a beehive retainer spring.

The countersunk gun set fits all sizes of flush head rivets. The countersunk rivet cannot use the beehive retainer spring because it will not fit over the gun set retainer ring. The countersunk ***rivet set*** uses a specially designed retainer spring. Figure 4-5 shows both types of retainer springs.

2. Bucking Bars

The tool used to form an ***upset head*** while using a pneumatic rivet gun is a bucking bar. Bucking bars are made in various shapes, sizes, and weights. The weight of the bucking bar must be proportional to the size of the rivet.

To obtain a proper upset head, a good technique to use is shown in Figure 4-6. As the gun is firing, press the bucking bar firmly against the forming rivet shank and roll the bar slightly. This rolling action will aid in the formation of a barrel-shaped bucktail. If the bucking bar is too light for the size of rivet and gun, the metal will bend toward the bucktail. If the bucking bar is not held firmly against the rivet shank, the metal will bend away from the gun.

The smooth face of the bucking bar must not be allowed to nick or scar. Nicks or scars on the face of the bucking bar will mar the bucktails and can lead to rivet failure. All scarred or marred bucktails must be drilled out and replaced with fresh, unmarked rivets.

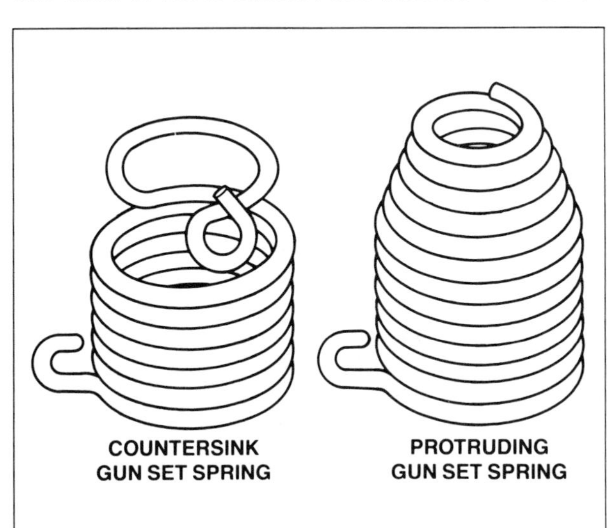

Fig. 4-5 Gun set retainer springs.

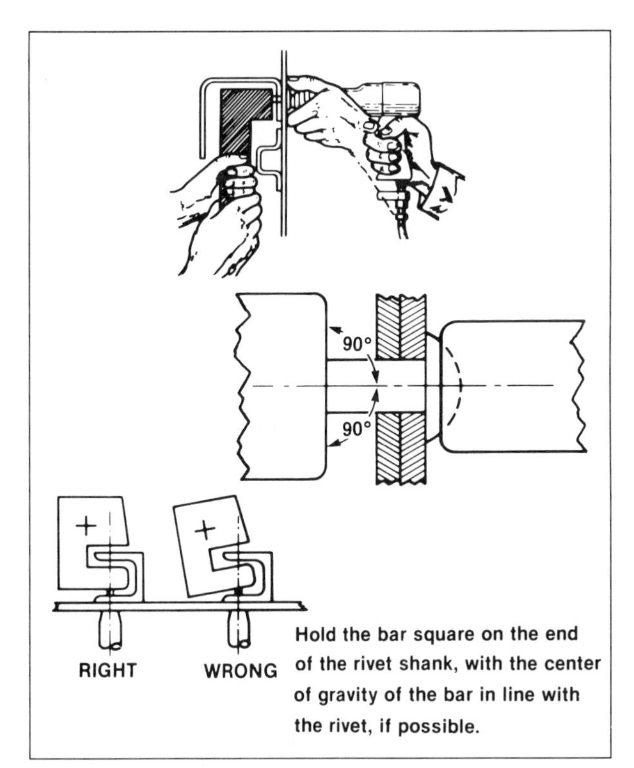

Fig. 4-6 Proper use of bucking bar.

B. Rivet Codes And Identification

The manufacturing of all solid-shank rivets is governed by Federal Specifications and Standards QQ-A-430. Solid-shank rivets are identified and cataloged by head shape, alloy content, shank diameter, and shank length (Figure 4-7).

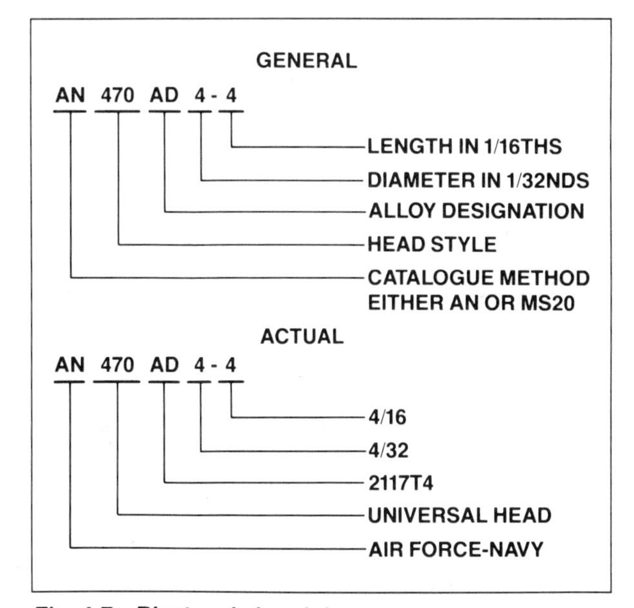

Fig. 4-7 Rivet code breakdown.

Two shorthand methods of coding are used to identify all aircraft rivets. An example is AN470AD4-5, or MS20470AD4-5. AN means Air Force-Navy and MS20 means Military Standards 20; 470 designates a universal style rivet head; AD refers to the alloy 2117T4; the hyphenated numbers designate the rivet diameter in thirty-seconds of an inch and the length in sixteenths of an inch. Thus, in the example, 4 means 4/32nds-inch diameter, and 5 means 5/16ths-inch length.

Although the AN and MS20 methods of cataloging rivets are similar, it is important to consistently use one of the two methods when ordering or identifying rivets. The rivet code is also used in blueprints, drawings, and technical manuals.

C. Rivet Head Styles

Four styles of rivet heads are used to construct an aircraft: Round, flat, universal, and countersunk. The latter two are the most commonly used in the aircraft industry. Figure 4-8 is a cross-sectional view of the universal and countersunk head rivets, showing their diameters and lengths.

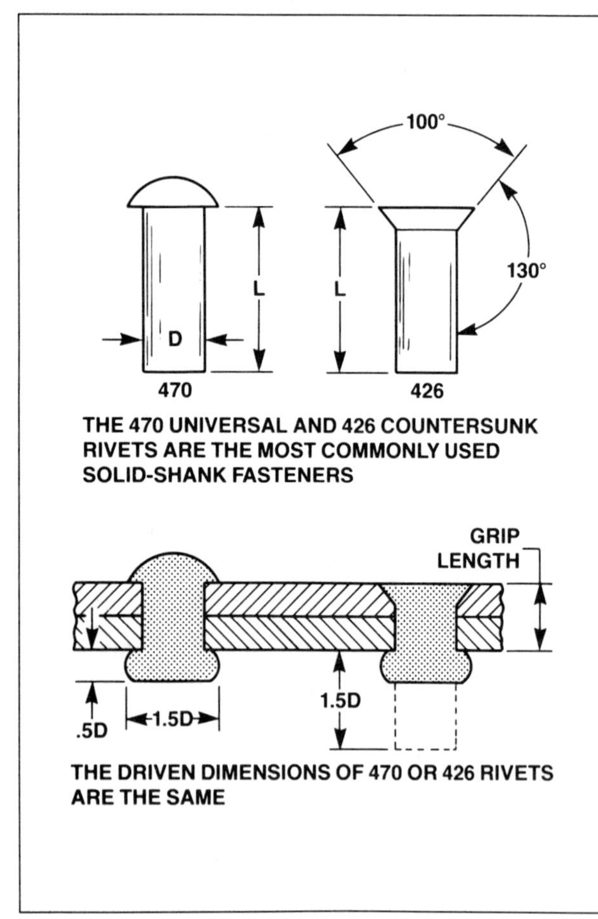

Fig. 4-8 Cross sectional view of universal and countersunk rivet heads.

1. Universal Head Rivets

Universal head rivets, also called ***protruding head rivets,*** are used internally in structural areas and on the skin surfaces of low to medium speed aircraft. A universal head rivet can withstand a much stronger ***bearing load*** than a countersunk rivet because the head is installed flat and binding on the surface of the riveted metal while the countersunk rivet is installed into a machine-tapered well.

The universal head rivet is a combination of several older head styles. When rivets were first used, various rivet head styles were available. Round head (AN430)rivets were used internally on high-strength structural areas. The flat head (AN442) rivet was used in tight areas where the round head could not be installed. Some modern jet aircraft still use round and flat head rivets in structural areas.

Aircraft built prior to World War II were low-speed aircraft, so a smooth aerodynamic air flow over the wing was not a major concern. As the speed of the aircraft increased, the need for smaller protruding head rivets accounted for the development of a modified brazier head (456), which causes less drag than larger protruding head rivets. Today, ***brazier head rivet*** styles are likely to be found only on aircraft built before 1955. Because the rivet sets used to drive rivets other than universals are difficult to obtain today, the older styles can be replaced by universal head rivets. Advisory Circular 43.13-1 explains the procedure.

2. Countersunk Rivets

As aircraft speeds increased, the need for smooth airfoils led to the development of the countersunk rivet. After experimenting with head angles of 78°, 90° and on high-speed jet fighters, 110°, the aircraft industry adopted a 100° standard. All of these experiments were attempts to increase the bearing strength of the rivet head around the skin.

The countersunk rivet has to be installed in a depression in such a way as to be flush with the surface of the skins it is holding together. The depression in the skin is called a nest or a well. The well can be made using a freehand or microstop countersink cutter.

Whenever the metal is cut to form a well or nest, the area around the rivet head is weakened. To compensate for this loss of strength, aircraft manufacturers must install a greater number of rivets in order to increase bearing and shearing strengths. Figure 4-9 shows how countersunk rivets are installed, by either machine or dimple methods.

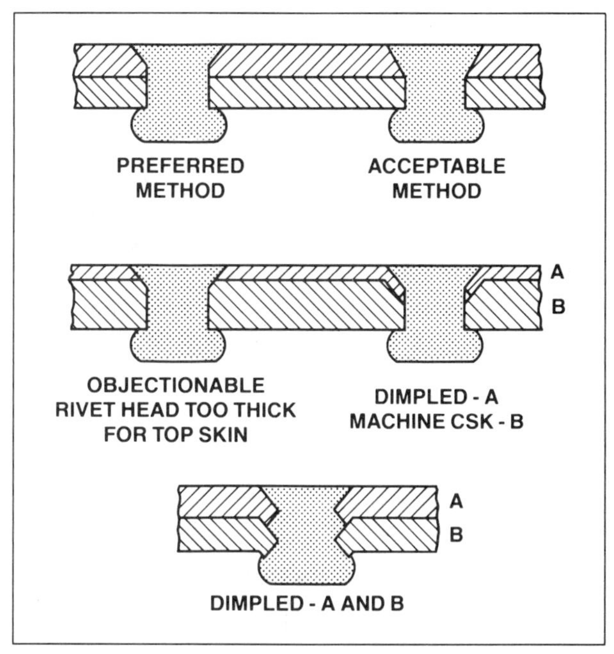

Fig. 4-9 Countersinking.

To remedy the loss in bearing strength caused by machine ***countersinking,*** the ***NACA*** (National Advisory Commission for Aeronautics) developed a method of countersinking that has been adopted by aircraft manufacturing companies. The NACA method of countersink riveting is illustrated in Figure 4-10.

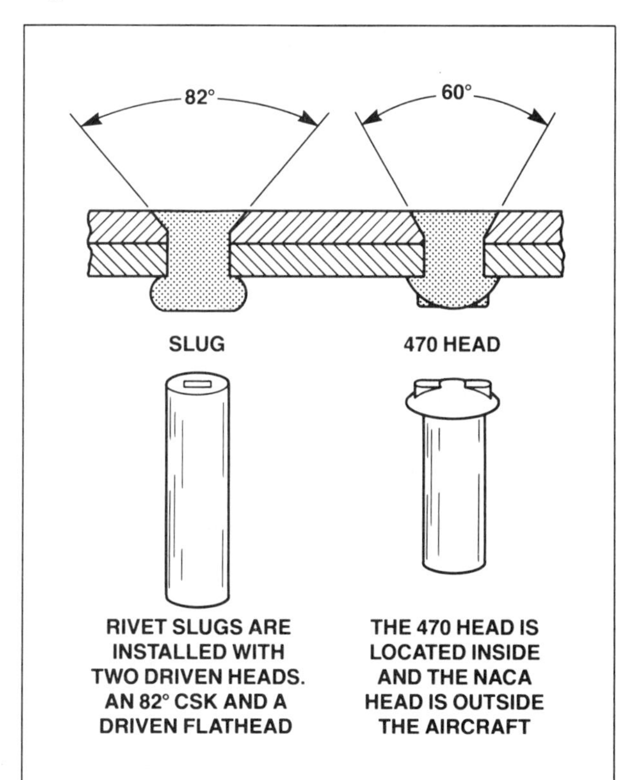

Fig. 4-10 NACA method.

Two different angles may be cut into the top skin, 60° or 82°. Military aircraft were the first to use the 60° well, on some of the older jet fighters. The 82° well is used when installing wing slugs on the Boeing 747. On some aircraft, a universal head is installed from the inside of the wing and driven into an 82° well.

Installing rivets using either the 60° or 82° NACA countersink method makes them as strong as universal head rivets. When coldworked, the bucktail formed in the 60° or 82° angle well is stronger than the conventional countersink riveting method because the driven head is packed into its well, creating a much stronger head than the regular countersunk rivet can produce.

D. Dimpling

Thin skins are never machine countersunk, because the cutter will go completely through the thin skin into the second skin and reduce the bearing strength around the countersunk rivet head. There is an alternative process called ***dimpling,*** which solves this problem.

Dimpling can be done in two ways, cold dimpling or hot dimpling. Cold dimpling of sheet metal skins is done on material less than .040 of an inch thick if countersunk rivets are required. The benefit of cold dimpling is that it produces stronger shearing and bearing strengths in the joint than would a driven universal head rivet of the same size.

Dimpling bars or sets can be made in the shop by cutting steel stock to the size needed for the job, drilling a hole the same size as the rivet to be used, and then setting a microstop countersink to cut about .015 of an inch deeper than the rivet head (Figure 4-11).

To use the dimpling bar, drill a hole into the sheet metal just as you would for a universal head rivet, place the dimpling bar under the rivet hole, and insert a countersink rivet. Using a rivet gun with a mushroom head set or a ball peen hammer, tap the rivet head into the dimpled well. Because dimpling does not produce the flushness of a machine countersunk rivet, be careful not to hit the area around the head too hard, or the metal surrounding the rivet will stretch, creating a problem which could be difficult to remedy. Metal that becomes stretched must be removed and replaced either by a patch or by changing a complete skin panel.

Hot dimpling is done to thicker aluminum alloys by pressing a well for the countersunk rivets. Thick aluminum alloy that is hot dimpled loses some of its heat treatment strength but, due to coldworking from driving the rivet, most of the skin strength is regained.

Hot dimpling consists of placing a predrilled hole over a female countersunk well and then pressing a male, heated die against it. The result is a coined dimple well embossed on the sheet metal part. A screw hole on an access plate is a good example of where a hot dimple technique might be used. The hot dimpling technique is frequently used by aircraft manufacturers and airline repair shops; the typical small-shop sheet metal mechanic may never have an opportunity to do hot dimple countersinking.

E. Rivet Alloys

Rivets used for the construction of aircraft must be as strong as, but much lighter than, steel. Alloying aluminum with other metals such as copper, magnesium, and zinc and then heat treating the metal makes this increase in strength possible. Alloyed aluminum rivets which are heat treated and allowed to cool and age harden are as strong as, and about one-third the weight of, steel. The alloys most commonly used to manufacture structural rivets are listed in Figure 4-12.

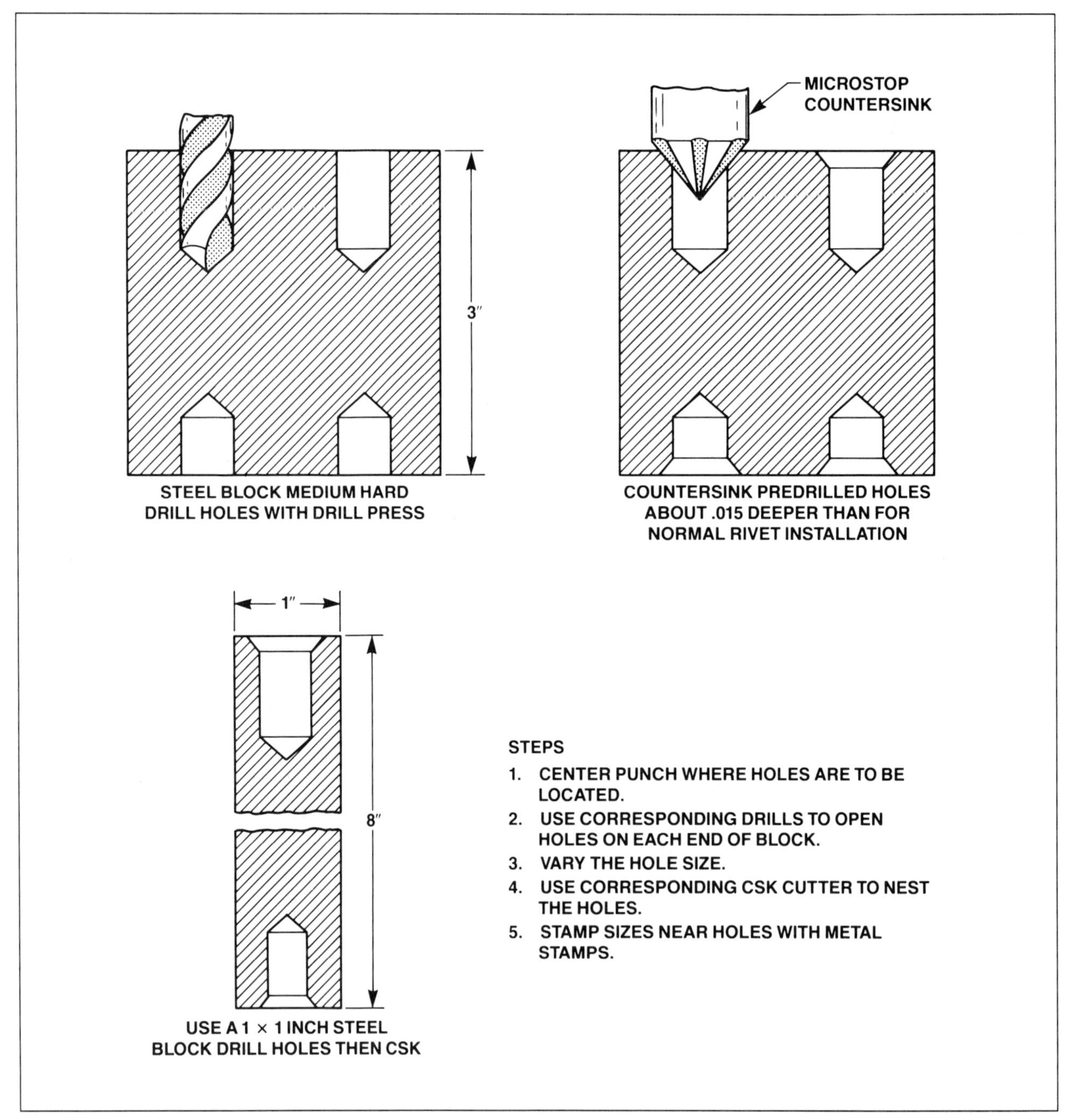

Fig. 4-11 Making dimple bars.

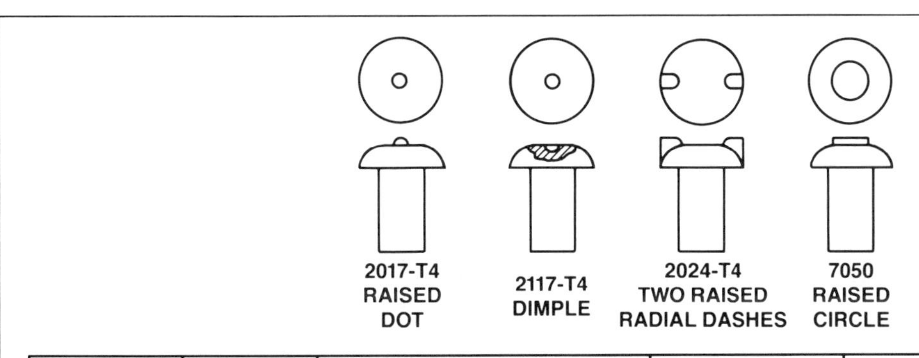

RIVET ALLOY	AN OR MS20 CODE	MARKING ON MFG HEAD	MAJOR ALLOYING INGREDIENT	CONDITION USED	AFTER DRIVING
1100	A	PLAIN	NONE	AS REC./MFG	1100
5056H32	B	RAISED CROSS	MAGNESIUM	AS REC./MFG	5056H32
2117T4	AD	DIMPLE	COPPER	AS REC./MFG	2117T3
2017T4	D*	RAISED DOT	COPPER	AS REC./MFG	2017T3
2017T4	D**	RAISED DOT	COPPER	RE-HEATED	2017T31
2024T4	DD	TWO RAISED SHOULDERS	COPPER	RE-HEATED	2024T31
7050T73	E	RAISED CIRCLE	ZINC	AS REC./MFG	7050T73

* MEANS ALCO CRACK FREE 2017T4 ALUMINUM ALLOY.
** MEANS RE-HEAT-TREATED ABOVE 3/16 INCH DIAMETER.

Fig. 4-12 Rivet alloy composition.

Rivet alloys 1100 and 5056H32 are non-heat-treatable. Non-heat-treatable rivets are rarely used on primary structures. The alloy 1100 contains an insignificant amount of copper (.01%). In the AN or MS20 methods of cataloging, the 1100 alloy is designated by the letter A. The 1100 rivet has no marking on the MFG (manufactured) head. These rivets are never used structurally, but are used in areas where strength is not a major concern.

The only non-heat-treated alloy used for structural work is 5056H32. It is used to skin magnesium-covered control surfaces because it is compatible with magnesium, its major alloying ingredient. Many light aircraft control surfaces were at one time covered with magnesium alloy to reduce weight. The only type of fastener that would not cause corrosion was the magnesium-compatible rivet 5056H32. The 5056H32 rivet is identified by a raised cross on its MFG head and it is designated by the letter B in the AN or MS20 catalog.

The alloy 2117T4, developed by Alcoa, is produced only as rivet stock. The 2117T4 rivet is called the "universal" replacement rivet, because it can be used in the "as received" condition from the rivet manufacturer. Rivets made from the alloy 2117T4 are more corrosion resistant and crack free than any of the other copper-bearing aluminum alloys. These rivets are used for plugging holes on wet wing fuel tank installations because of their excellent swelling characteristics. The 2117T4 rivet is designated by the letters AD in the AN and MS20 system of coding, and it has an identifying dimple embossed in the center of its MFG head.

The alloy 2017T4 is presently made in two forms of rivet stock. The first type, used in the construction of earlier aircraft, is slightly harder than 2117T4, and it is difficult to control its cracking when rivets of diameters greater than 3/16 of an inch are driven. To minimize cracking, such rivets must be re-heat treated before driving.

A new form, the crack-free 2017T4 rivet alloy, recently developed by Alcoa, was first used for skinning the Boeing 727 wing. It contains slightly less magnesium and a precisely controlled amount of iron and silicon. The new crack-free alloy has a greater shear strength (38KSI) than the older form (34KSI). The two forms of 2017T4 are designated by the letter D and have a raised dot on their MFG heads.

The alloy 2024T4 is produced as skins, extrusions, and rivets. The 2024T3 rivet is used in the construction of light and heavy aircraft, both internally and externally, and has long been established as the strongest of the copper-bearing aluminum alloys. Rivets made from alloy 2024 cannot be driven without first softening them by re-heat treating. The alloy 2024T4 is denoted by the letters DD and has two raised shoulders on its MFG head.

In 1979, Alcoa and the US Air Force developed a new rivet alloy, 7050T73. The alloy 7050 in a stable T73 condition is a rivet material with good formability, high strength, and excellent resistance to corrosion cracking. The material used to make the alloy 7050 is a mixture of magnesium and zinc. This alloy has a fatigue life limit longer than that of the aircraft structure itself. Rivets of 7050 alloy in the T73 temper are considered to be the strongest aluminum rivets commercially available, and they should be considered as replacements for 2024T31 rivets on new aircraft designs.

For example, the Boeing 767 uses the 7050T73 rivet extensively. In addition, its ability to be driven in the "as received" condition makes the Alcoa 7050T73 alloy rivet an excellent choice for automatic rivet setting and field or shop repair work. The 7050T73 rivet is identified by a ring on its MFG head and is coded by the letter E.

The 7050T73 alloy is available in all head styles. The countersunk, counterbored, tension-type rivet (Figure 4-13) made from the alloy was developed to overcome the loss of strength which results from a conventional machine-type countersink job. One of the primary uses of this rivet has been to skin wings and tail surfaces of modern jet aircraft.

1. Icebox Rivets

"Icebox" is a nickname given to rivets made of alloy 2017T4 more than 3/16 of an inch in diameter and all rivets of the alloy 2024T4. These two alloys are too hard to drive in their original heat-treated condition. They must be re-heat treated and kept in a frozen condition until ready for use.

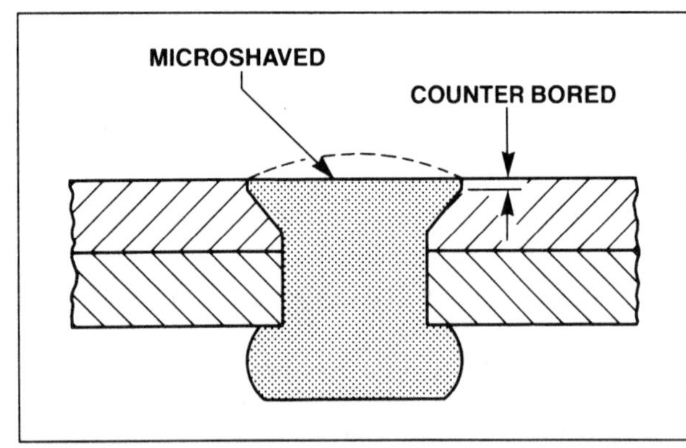

Fig. 4-13 Countersink counterbored tension-type rivet. Counterbore must be made with a precision cutter. This rivet is stronger than conventional CSK rivet because MFG head is packed into the counterbored well, giving area additional tension or head bearing strength.

When these two alloys are heat-treated, they harden after cooling down, due to a process called age hardening. Age hardening brings the metal to its maximum natural hardness, which can be held in suspension by storing these rivets in a freezer chest. Years ago when this process was first developed, there were no freezers, so the rivets were stored in dry ice boxes; thus the name ***icebox rivets.***

Today, the re-heating process is the same as the one originally used to heat treat the rivet, but immediately after the rivet is quenched, it is put into identified containers and stored in a freezer. The newer nickname could be "freezer" rivets.

Alloys such as 2024 are followed by a temper designation, made up of a letter and a number, which indicates the temper condition of the rivet. T4 means solution heat treated and age hardened. T3 means solution heat treated, age hardened, and then cold-worked. A T3 rivet is one which has been driven. All driven rivets in the 2XXX series alloys have a T3 or T31 designation. When a rivet like 2024 has the temper designation T31, as in 2024T31, it means that the rivet is an icebox rivet. All driven icebox rivets have the T31 tag added to the alloy code.

For a long time, the 2024T31 icebox rivet was considered to be the strongest alloy. However, the new rivet 7050T73 may replace the 2024T31 icebox rivet. The only reason it has not yet done so is that it is presently more expensive. But if one considers the cost of the icebox step, the cost of the 7050 rivet is comparable. In addition, the ***FAA*** may soon approve the 7050T73 rivet as a replacement for the 2024T31 rivet for general repair work, as it has already approved its use on certain types of airline equipment.

2. Rivet Replacement Rules

The correct alloy must be used in making structural repairs or rivet replacements because the strength of the repair depends upon the total shear and bearing strength of the joint. As a general rule, rivets are replaced by rivets of the same alloy and size. However, 2024 rivets may be replaced by oversized rivets of lower strength. See Advisory Circular 43.13-1, paragraph 99e.

If the replacement rivet is larger than the maximum specified size, another rule applies: "Rivets may not be replaced by a type having lower strength properties, unless the lower strength is adequately compensated by an increase in size or a greater number of rivets" (paragraph 99c).

For example, twenty 2024T31 rivets can be replaced by twenty-eight 2117T3 rivets. Since each 2024T31 rivet has a driven shear strength of 1175 pounds, the total shear strength is 23,500 pounds,

20 × 1175. Since each 2117T3 rivet has a driven shear strength of 860 pounds, the total shear strength of the replacement is 24,080 pounds, 28 × 860. Thus, the replacement, using eight more rivets of lesser strength, has 580 pounds more shear strength than the original.

F. Shear Strength

Shear strength is the ability of the rivet material to withstand being cut in half. The power to resist this cutting action is called ultimate shear strength. The shear strength of a riveted joint depends upon the total number of rivets used to make the joint. In computing the shear strength of a driven rivet in pounds, use the drill hole size, not the nominal hole size. Nominal hole size is the same size as the rivet diameter. Drill hole size is .002 to .004 of an inch larger than the rivet diameter.

Alcoa developed a shear strength formula (Figure 4-14) which can be used to derive the shear values in Figure 4-15. For example, an AN470AD4 rivet has a driven shear strength of 389 pounds. The minimum shear strength for the alloy 2117T3 is 30,000 PSI. The area for a #30 drill is .01296 square inches, as is found in Figure 4-16.

Driven shear strength = 30,000 × .01296 × 1 = 389 pounds of shear.

(See appendix B for a program in IBM basic language that can be used to compute shear strength.)

In Figure 4-15, the shear strengths are in pounds and in Figure 4-17 they are in *KSI*. 1,000 PSI equals 1 KSI. For example, 30 KSI is equal to 30,000 PSI.

Shear strength, although it is independent of bearing strength, is affected by its load as the joint is worked. Ideally, shear and bearing strengths should be nearly the same. But they are not, because as metal gets thicker, the joint moves from high bearing to high shearing loads.

Any deformation to the rivet shank is classified as shear failure. For example, a rivet cut in half or joggled is shear failure. Shear failure can be remedied by adding larger diameter rivets and doubler plates to absorb the load.

G. Bearing Failure

Bearing strength is the ability of the metal to withstand being torn or ripped away from the rivets or riveted joint area. If the metal tears away from the rivets, the joint is said to have failed in bearing strength. Bearing loads are most critical on thin skins because the skin cannot absorb the swelling process of the solid-shank rivet as well as thick skins can. On some joints, the skin will stretch and bulge around the rivet heads. A remedy for stretching is to place a metal doubler under the primary skin joint to absorb the hitting force of the rivet gun.

Bearing strength problems often occur in joints made with countersunk rivets, because strength is lost when the metal is shaved away to make the rivet well.

ULTIMATE SHEAR LOAD = $F_S \times A_S \times N$

F_S IS THE DRIVEN SHEAR STRENGTH IN PSI OR KSI OF A RIVET.

A_S IS THE AREA OF A RIVET HOLE. THE HOLE SIZE IS EQUAL TO THE DRILL SIZE.

N IS THE NUMBER OF SHEARING PLANES.

FOR EXAMPLE, THE SHEAR STRENGTH OF A DRIVEN AN470AD4 RIVET IS COMPUTED AS FOLLOWS: USE KSI FROM FIGURE 4-15 AND HOLE AREA FROM FIGURE 4-16.

USL = 30,000 × .01296 × 1 = 388.8 OR 389 POUNDS OF SHEAR.

Fig. 4-14 Shear strength formula.

ALLOYS	DRIVEN SHEAR STRENGTH IN KSI F_S	RIVET DIAMETER						
		3/32	1/8	5/32	3/16	1/4	5/16	3/8
2117T3	30	217	389	596	860	1560	2460	3510
2017T3	38	275	492	755	1089	1971	3114	—
2024T31	41	297	531	814	1175	2127	3360	—
5056H32	28	203	363	566	802	1452	2294	—
7050T73	43	311	557	854	1230	2230	3520	5030

Fig. 4-15 Single shear strength of aluminum alloy rivets.

NOMINAL RIVET DIAMETER, IN.		1/16	3/32	1/8	5/32	3/16	1/4	5/16	3/8
RECOMMENDED HOLE DIAMETER, IN.		0.067	0.096	0.1285	0.159	0.191	0.257	0.323	0.386
CORRESPONDING DRILL SIZE		51	41	30	21	11	F	P	W
CORRESPONDING SINGLE SHEAR AREA, SQ. IN (A_S)		0.00353	0.00724	0.01296	0.01986	0.02865	0.05187	0.08194	0.1170
BEARING AREA, SQ IN.									
SHEET AND PLATE THICKNESSES, IN.	0.016	0.00107	—	—	—	—	—	—	—
	0.020	0.00134	0.00192	—	—	—	—	—	—
	0.025	0.00168	0.00240	0.00321	—	—	—	—	—
	0.032	0.00214	0.00307	0.00411	0.00509	—	—	—	—
	0.040	0.00268	0.00384	0.00514	0.00636	0.00764	—	—	—
	0.050	0.00335	0.00480	0.00643	0.00795	0.00955	0.01285	—	—
	0.063	0.00422	0.00605	0.00810	0.01002	0.01203	0.01619	0.0204	—
	0.071	0.00476	0.00682	0.00912	0.01129	0.01356	0.01825	0.0229	0.0274
	0.080	—	0.00768	0.01028	0.01272	0.01528	0.02056	0.0258	0.0309
	0.090	—	0.00864	0.01157	0.01431	0.01719	0.02313	0.0291	0.0347
	0.100	—	0.00960	0.01285	0.01590	0.01910	0.02570	0.0323	0.0386
	0.125	—	—	0.01606	0.01988	0.02388	0.03213	0.0404	0.0483
	0.160	—	—	—	0.02544	0.03056	0.04112	0.0517	0.0618
	0.190	—	—	—	—	0.03629	0.04883	0.0614	0.0733
	1/4	—	—	—	—	—	0.06425	0.0808	0.0965
	5/16	—	—	—	—	—	—	0.1009	0.1206
	3/8	—	—	—	—	—	—	—	0.1448
	1/2	—	—	—	—	—	—	—	—
	5/8	—	—	—	—	—	—	—	—
	3/4	—	—	—	—	—	—	—	—
	1	—	—	—	—	—	—	—	—

Fig. 4-16 Recommended hole sizes for cold-driven solid rivets with corresponding shear and bearing areas.

ALLOY AND TEMPER BEFORE DRIVING	MINIMUM UNDRIVEN SHEAR STRENGTH, KSI	DRIVING PROCEDURE	TEMPER DESIGNATION AFTER DRIVING	DRIVEN SHEAR STRENGTH, KSI
2017-T4	33	COLD, AS RECEIVED	T3	38
2017-T4	33	COLD, IMMEDIATELY AFTER QUENCHING OR REFRIGERATED	T31	34
2024-T4	37	COLD, IMMEDIATELY AFTER QUENCHING OR REFRIGERATED	T31	41
2117-T4	26	COLD, AS RECEIVED	T3	30
2219-T81	32	COLD, AS RECEIVED	T81	36
5057-H32	24	COLD, AS RECEIVED	H321	28
7050-T73	41	COLD, AS RECEIVED	T73	43

Fig. 4-17 Shear strengths in KSI.

This could be prevented by hot or cold dimpling. When metal is machine countersunk, it is physically cut away from the thickness of the top skin and the area is weakened. In addition, a machine countersunk 426 rivet head does not get coldworked like a universal head rivet because the mushroom gun set contacts only the flat portion of the rivet head.

Thus, energy which would normally coldwork a protruding head rivet is lost to the surrounding skin area. This is not entirely bad because the skin will assume part of the lost countersunk head strength. However, in most cases, the metal becomes stretched and dished.

In short, machine countersunk rivets are not as strong as protruding head rivets, and when they are used to streamline a wing or control surface, more rivets must be used to make up for the loss of bearing strength.

Alcoa Crown ***Flush Rivet*** can be countersunk without endangering the bearing strength of the rivet head. The top of the head of the Crown Flush is in contact with the mushroom gun set. This action causes the gun to drive the countersunk head into the well, and at the same time expansion of the shank is completed. The result is a fully coldworked rivet driven into a well that needs no microshaving (Figure 4-18).

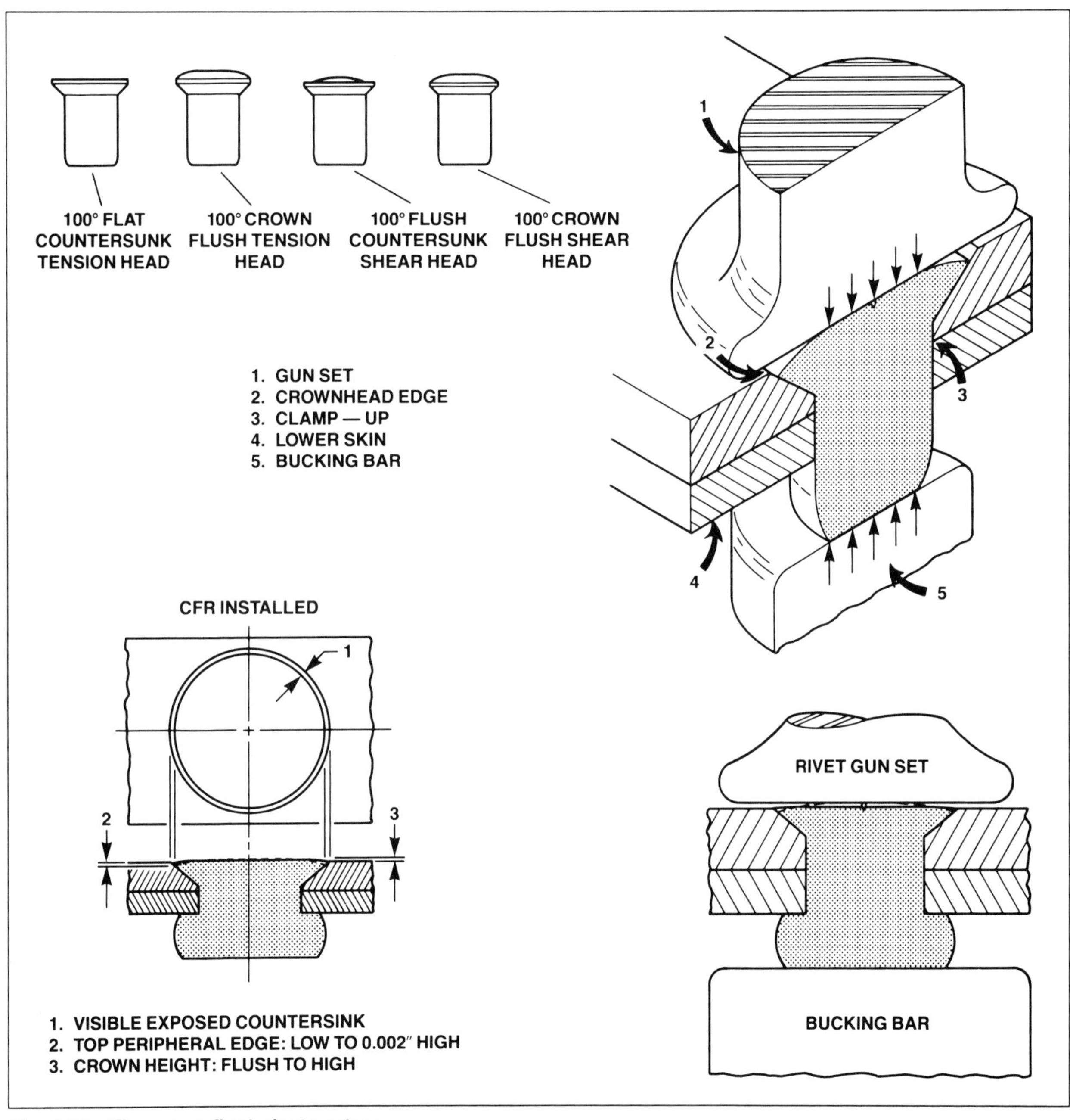

Fig. 4-18 The crown flush rivet system.

H. Inspection Of A Rivet Joint

There are three places to check when inspecting a rivet joint: The MFG head, the shop head, and the skin around the rivet heads. Any damage to either of the two rivet heads is not critical because rivets can be drilled out and replaced.

Note: Never oversize the hole when drilling out a damaged rivet.

If skin damage is extensive, a new skin panel must be installed.

The procedures for removing rivets depend upon the situation. If, for example, an aircraft has extensive damage to a wing leading edge or a section of the spar, the mechanic will remove the damaged metal as well as some of the parts that are still usable. In removing rivets from reusable aircraft parts, it is essential that the rivet holes not be oversized.

The correct way to remove the rivets is to file a flat spot on all protruding head rivets except the 2117 rivets, to center punch each MFG head, to back up each rivet with a bucking bar, to select a drill ONE SIZE SMALLER THAN THE RIVET BEING REMOVED, to drill only the depth of the MFG head, to use a pin punch which is the same size as the drill, to snap off the drilled MFG heads, and to back up each remaining stem by tapping out the shank without stretching the metal.

A different procedure is followed in removing the occasional rivet badly driven during reassembly. Such rivets should be removed by the same size drill as the rivets being installed. Drill the depth of the MFG head only; then lightly tap off the MFG head and gently knock out the remaining shank.

Figure 4-19 illustrates an assortment of faulty rivets which must be removed and replaced. The most troublesome rivet fault is a clinched rivet, which results from improper bucking action. The rivet forms to one side, which can lead to a corrosive condition at a later date. Rivets that crack do so because they became too hard while the bucktail was forming. This is a result of hitting the rivet too lightly or allowing an icebox rivet to recover its age hardening by keeping it out of the freezer too long before driving.

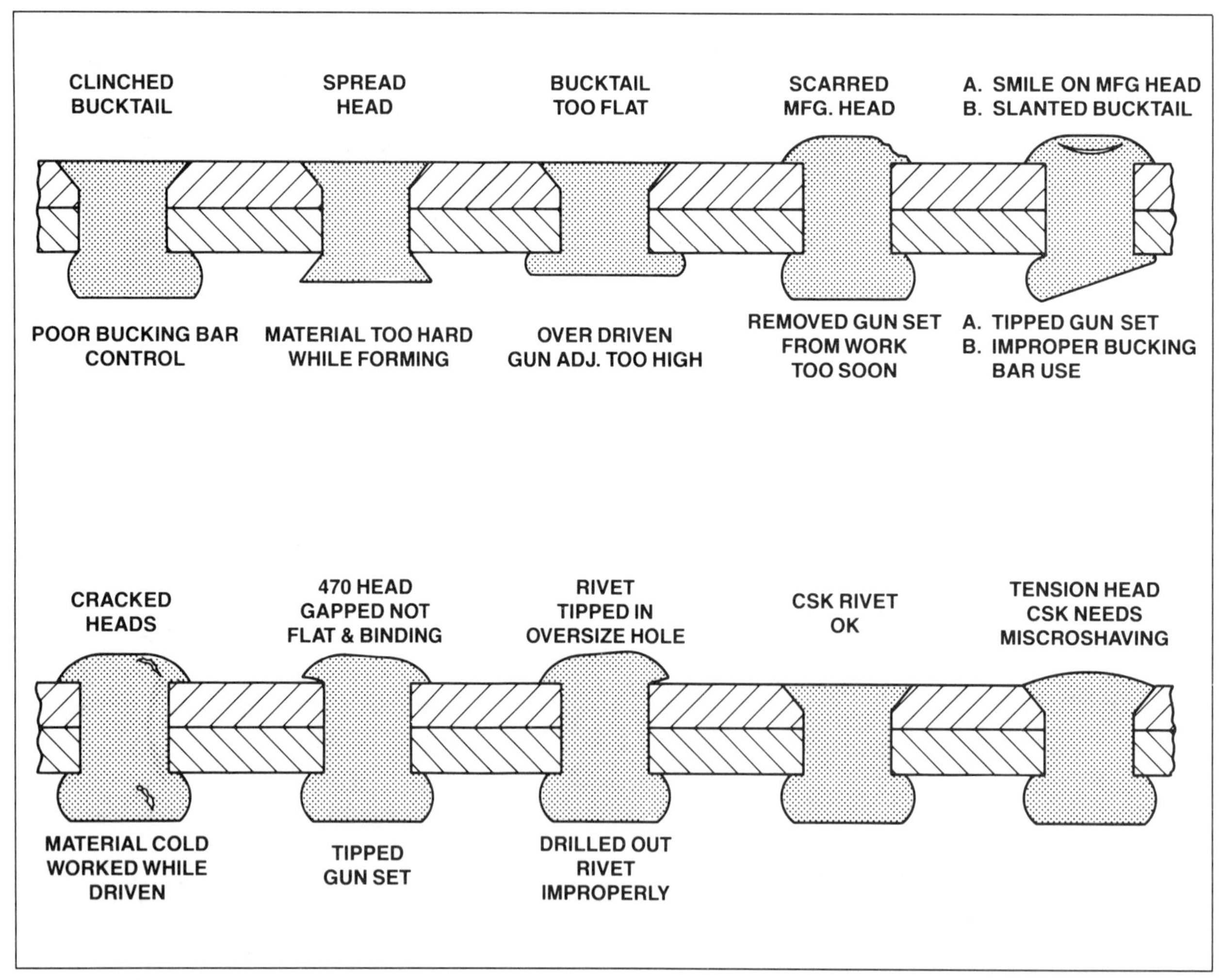

Fig. 4-19 Rivet faults and reasons.

I. Selecting Rivet Diameters And Lengths

In order to determine the correct rivet diameter, refer to the manufacturer's manuals, compare the area to be riveted with another aircraft of the same model and type, or multiply the thickness of the skin, spar, or longeron by three and then select a 1/32-inch larger rivet for the job. See appendix B for an IBM Basic Language computer program which can be used to find the rivet diameter.

Correct rivet length depends upon the thickness of the metal (grip length) and the diameter of the rivet. To find the length of the rivet, add grip length to the diameter of the rivet multiplied by 1.5. Then select the nearest larger size in 1/16ths. Cut the rivets to their driving size with a pair of rivet cutters.

QUESTIONS:

1. *Name four kinds of heat-treatable aluminum alloy rivets used in manufacturing aircraft.*
2. *What are the Cleco colors that identify 3/32-, 1/8-, 5/32- and 3/16-inch diameter rivets?*
3. *What size drill is used to open holes for a 1/8"D rivet?*
4. *What is the name of the Alcoa rivet alloy which can replace icebox rivets?*
5. *What aluminum alloy rivet is best suited for riveting magnesium skins?*
6. *What marking does the rivet alloy 2117T4 have on the MFG head?*
7. *Why do icebox rivets have to be reheated and/or refrigerated before driving?*
8. *What is the name of the Alcoa alloy that was once considered an icebox rivet?*
9. *Convert 34 KSI to PSI.*
10. *What is the AN letter code for the alloy 2024T4?*
11. *What joint strength is weakened by the use of machine countersunk rivets?*
12. *Describe the NACA method of countersinking rivets.*
13. *All thicknesses being equal, which rivet joint would have the highest shear strength: Universal head rivets, machine countersunk rivets, or cold dimple countersunk rivets?*
14. *What is the maximum allowable edge distance for aircraft riveting?*
15. *What is the formula for finding driven shear strength?*

Chapter V
Aircraft Special Fasteners

Aircraft special fasteners are divided into two main categories: conventional and blind fasteners. Special fasteners are so named because they often require special tools for installation and they are made to meet specific structural needs. Blind fasteners are made to be used in locations where the conventional solid-shank rivet and aircraft nuts and bolts cannot be accessed for installation.

There have been many changes in fasteners over the last twenty years due to the increased speed and weight of modern jet aircraft. More than 30,000 different types of fasteners are now manufactured, many of which are used on modern aircraft. This chapter is a review, necessarily selective, of the fasteners most commonly used on civilian aircraft, and an introduction to some new types of fasteners available for the construction and repair of aircraft.

A. Solid-Shank Rivets

Solid-shank rivets are used in the construction of wide-body jetliners and light aircraft. But solid-shank rivets are limited in their use because they lose their excellent loading characteristics when larger diameters and greater strength are required. Because most solid-shank rivets are made from aluminum alloy, they do not have the strength required for large jet aircraft. The diameters of such rivets would have to be very large to meet necessary bearing and shearing strength requirements. Beyond a certain size, then, solid-shank rivets are impractical to use.

B. Hi-Shear Rivets

Hi-Shear rivets, first used during World War II to construct heavy bombers, were developed to withstand the high ***shear loads*** of larger aircraft.

The Hi-Shear rivet is available in two styles: conventional and blind. The conventional Hi-Shear fastener has two main parts, a steel stud and a crushable collar made of either lead or 2117T4 aluminum. The steel stud, as strong as a standard AN bolt, is held in place by the collar. Hi-Shear rivets are available in two head styles — countersunk and flat. The diameters of Hi-Shear rivets are measured in 1/32-inch increments. Hi-Shear rivet lengths are measured in 1/16-inch increments.

The installation tool used for Hi-Shear rivets is either a pneudraulic ***squeeze gun*** or a conventional pneumatic rivet hammer with a specially designed gun set that fits over the collar to form a cone. The installation of Hi-Shear rivets (Figure 5-1) is typical of all straight-pin type special fasteners, those whose shanks are straight and which look like standard AN bolts. These installations require close tolerance holes. That is, the rivet is forced or tapped into the hole, making the fastener fit tightly and leaving no room for joint slippage.

When installing special fasteners, the aircraft manufacturer specifies the hole clearance which will allow for the best joint strength. For example, if the required hole is 1/4-inch in diameter, the manufacturer specifies drilling the hole .007 inch undersize and then bringing it up to size using a reamer. If the Hi-Shear stud is too long, the collar

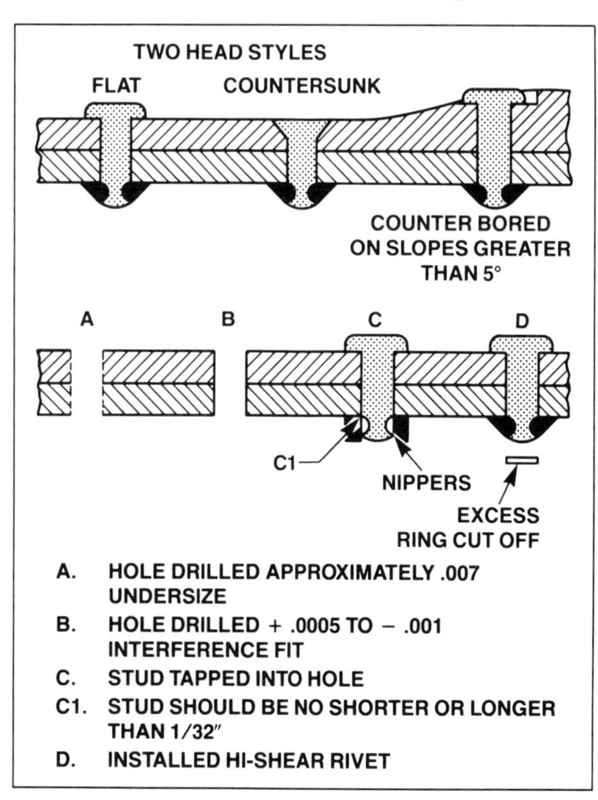

Fig. 5-1 Hi-Shear rivets.

will not form a full locking cone in the space provided and the bearing strength of the collar will be weakened. If the stud is too long, plain washers may be used under the collar or MFG (manufacturer's) head.

The correct way to fit the length of a Hi-Shear rivet is to insert a specially made, slightly smaller diameter Hi-Shear rivet into the hole. The straight portion of the shank should not extend more than 1/32-inch through the metal. The Hi-Shear rivet has a sharp edge on the lower part of the stud that nips off the excess amount of collar in the form of a discarded ring.

Hi-Shear rivets cannot be used on a surface slanted more than 5° unless the surface is ***spot-faced.*** A spotfacing tool is similar to a large counterbore and is used to make a flat surface for the Hi-Shear collar or head to bear on. The cut made by the spotfacing tool should be large enough for the Hi-Shear installation tool to fit into the well. After a Hi-Shear rivet is installed into a spotfaced well, the fastener and bearing surface must be coated with zinc chromate primer.

1. Blind Hi-Shear Fasteners

The Hi-Shear rivet is also available as a blind fastener. Preassembled at the factory, it consists of a threaded steel alloy bolt, a threaded steel nut with a serrated edge, and an expandable steel sleeve. The blind Hi-Shear fastener is available in four diameters — 3/16, 1/4, 5/16 and 3/8-inch, and in three head styles — flush, hex-head, and flush millable. The newer types of blind Hi-Shear bolts are made from a variety of metal alloys, such as stainless steel, titanium, and aluminum.

Figure 5-2 shows the blind Hi-Shear bolt preinstalled and installed. As the blind Hi-Shear fastener is being installed, a bolt is turned inside of a serrated edge nut. The serrated nut locks against the walls of the hollow rivet shank, preventing it from turning while the ***clamping-up*** action is taking place. This causes the sleeve to expand over the end of the nut, forming an expanded blind head which clamps against the work. Installations of blind fasteners in primary aircraft structures usually require an inspector's approval.

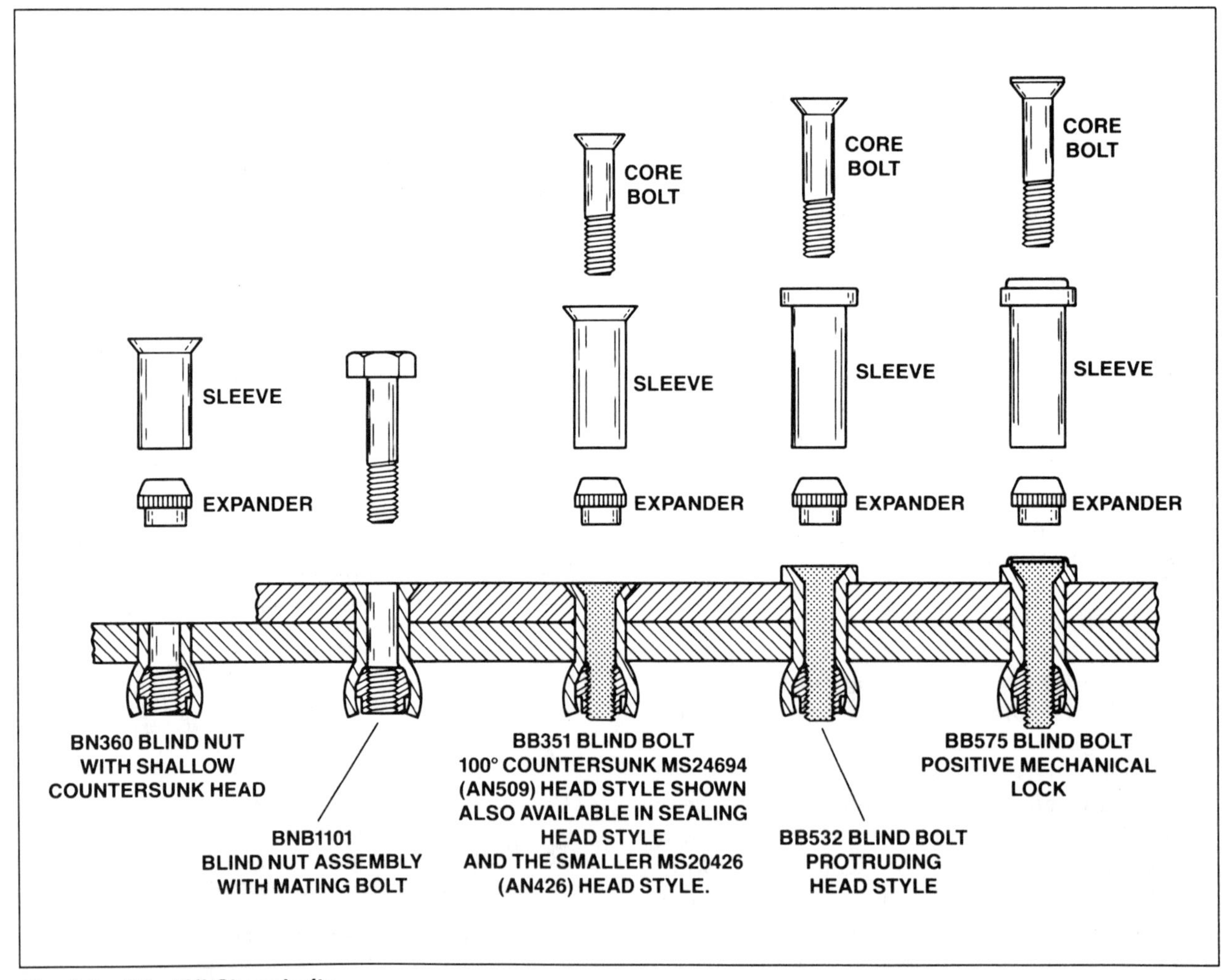

Fig. 5-2 Blind Hi-Shear bolts.

C. Hi-Lok Fasteners

Hi-Loks are made in several different alloys, such as aluminum alloy, steel, stainless steel, and titanium. They have two head styles, flat and countersunk. Hi-Loks are installed in the same manner as Hi-Shear rivets with regard to interference fit. The tools needed to install these fasteners are a ratchet with a hole drilled through the socket holder and an Allen wrench that fits through ratchet hole and up into the Hi-Lok shank. The locking device on the Hi-Lok is a collar with a shear nut attached to it.

The installation of the locking collar on to the stud is shown in Figure 5-3. The shear nut connects to the locking collar and shears away when the proper torque is reached. The amount of torque required to shear the collar nut is not enough to stretch and to shrink its shank, as in the case of a standard AN nut and bolt assembly. This feature makes the fastener valuable structurally because its diameter does not decrease when it is installed. The Hi-Lok has a threaded collar which is compatible with the fastener, permitting it to withstand high bearing and shearing loads.

1. Hi-Lok Hi-Tigue Fastening System

The Hi-Lok Hi-Tigue is a prestressing fastener system which does not require a tapered hole. The hole preparation and the final installation of the Hi-Lok Hi-Tigue fastener are shown in Figure 5-4.

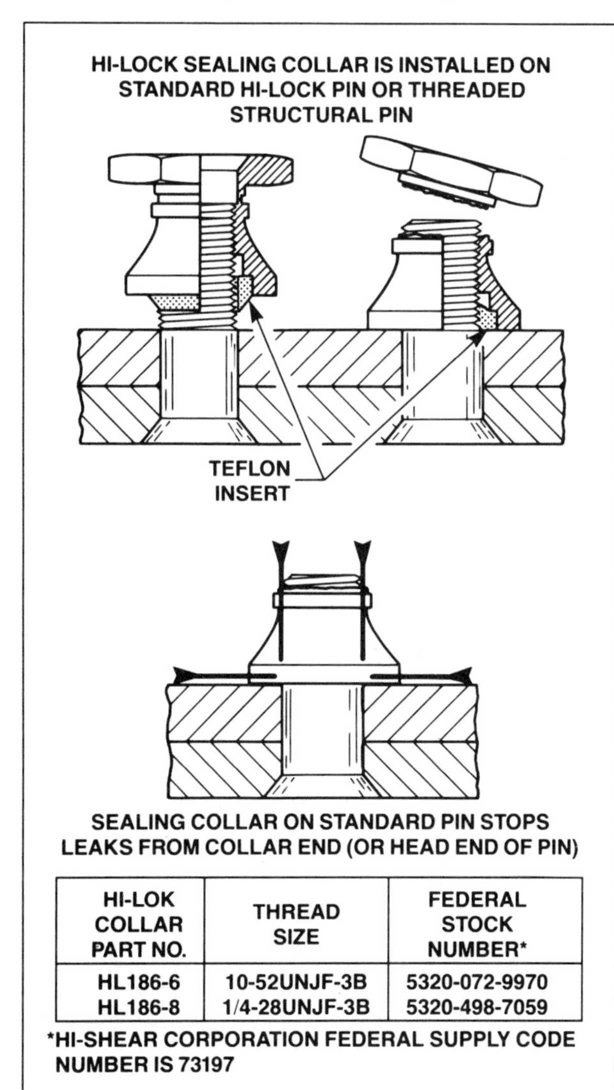

HI-LOK COLLAR PART NO.	THREAD SIZE	FEDERAL STOCK NUMBER*
HL186-6	10-52UNJF-3B	5320-072-9970
HL186-8	1/4-28UNJF-3B	5320-498-7059

*HI-SHEAR CORPORATION FEDERAL SUPPLY CODE NUMBER IS 73197

Fig. 5-3 Hi-Lok sealing collars.

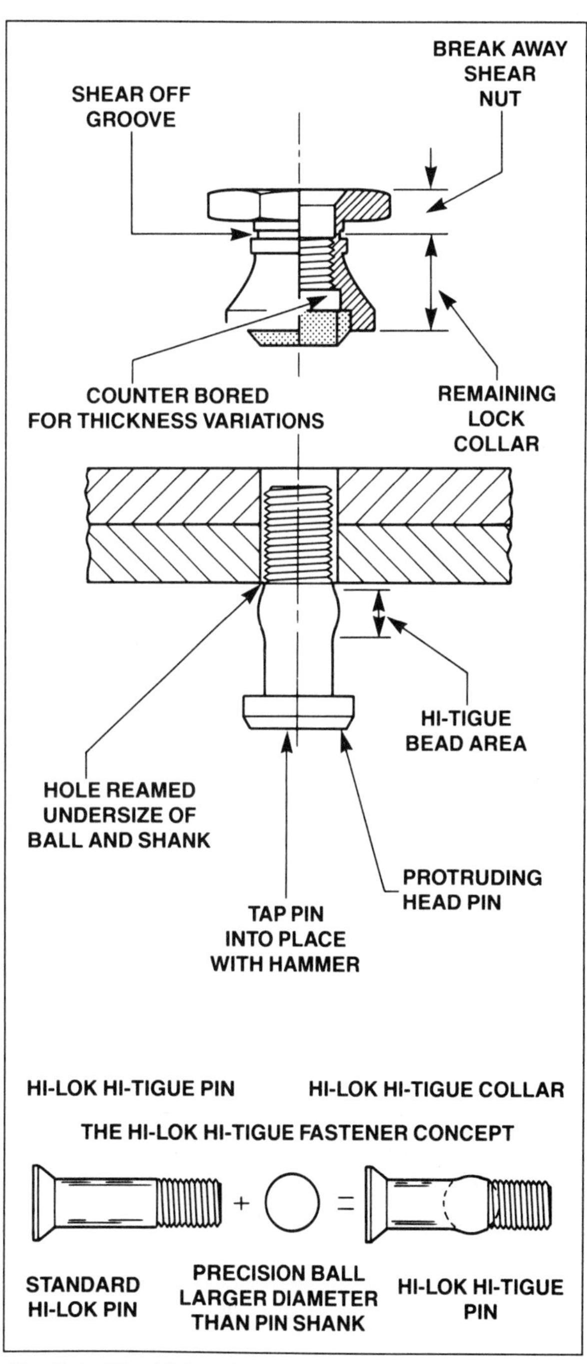

Fig. 5-4 The Hi-Lok Hi-Tigue Fastener system.

The hole has to be prepared slightly undersize of the shank in order to get what the manufacturer calls the ***ballizing*** effect. After installation, the sliding ball generates a plastic action, thereby allowing the metal wall behind the ball to become burnished and causing an elastic recovery of the hole surface behind the bead of the Hi-Tigue pin. The ballizing section of the Hi-Tigue pin creates a radial force against the wall of the hole which, in turn, preloads the area.

Figure 5-5, a load and deformation chart, shows the stress and strain curve of the cylindrical hole, which recovers part of its elasticity due to the deformation of the metal.

The Hi-Tigue fastener can be produced in any of the popular weight-saving, high-strength and high-temperature materials such as titanium alloys, alloy steel and stainless steel. Sizes range from 1/8-inch through one-inch diameters in increments of 1/32-inch. A wide range of grip lengths, measured in 1/16-inch increments, is available. The head styles include shear or ***tension*** types of protruding or 100 flush heads.

The collars of the Hi-Tigue fastener are of two types: standard and sealing. The standard collars are styled for tension and shear application. The sealing collar is used for wet wing and tank areas. The sealing collar has a Teflon sealing insert which is able to withstand temperatures of 500°F (270°C). Sealing agents other than Teflon are also available for this collar.

Because the Hi-Lok Hi-Tigue fastener can be installed with hand tools, it can be used in general aviation shops. The Hi-Tigue pin can be pressed into the interference hole by tapping it in with a mallet. After it is inserted, a compatible collar is installed using a socket with a special ratchet and an ordinary Allen wrench (Figure 5-6). The torque-off feature of the nut on the collar preloads the stack-up of the metal parts being joined.

D. Hi-Lite Fastener

The Hi-Shear Hi-Lite is a new aircraft fastener. The Hi-Lite, as the name implies, saves weight. Because its shank-to-thread transition is shorter than that of conventional threaded fasteners (Figure 5-7), it has high fatigue strength. In addition, its stronger collar allows the pin threads to be shortened, thus saving weight without sacrificing strength.

Hi-Lites are available in an assortment of diameters ranging from 3/16 to 3/8-inch. They are available in lengths measured in 1/16-inch increments.

Four locking collars are used to hold the Hi-Lite into position: Hi-Lite multi-purpose ST pin, swaged collar, forged collar, and HCL ST collar. All but the swaged collar, which has to be pressed onto the threaded portion of the Hi-Lite shank, are threaded like the Hi-Lok. The HCL ST collar is used on honeycomb surfaces.

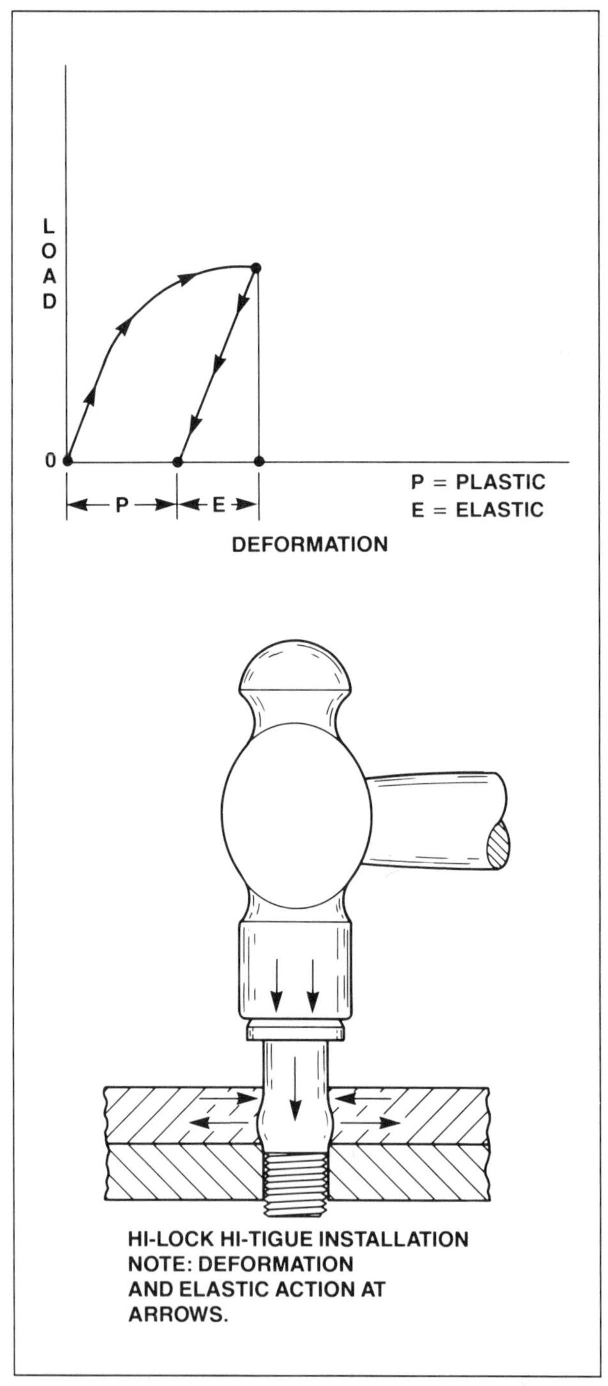

Fig. 5-5 Stress and strain curve of Hi-Lok Hi-Tigue. Chart shows elastic recovery of Hi-Lok Hi-Tigue pin when forced into undersized hole. Amount of hole recovery is less because of deformation. The part (E) which does recover makes cylindrical wall surface stronger and harder. Action of ball causes area to become pre-loaded.

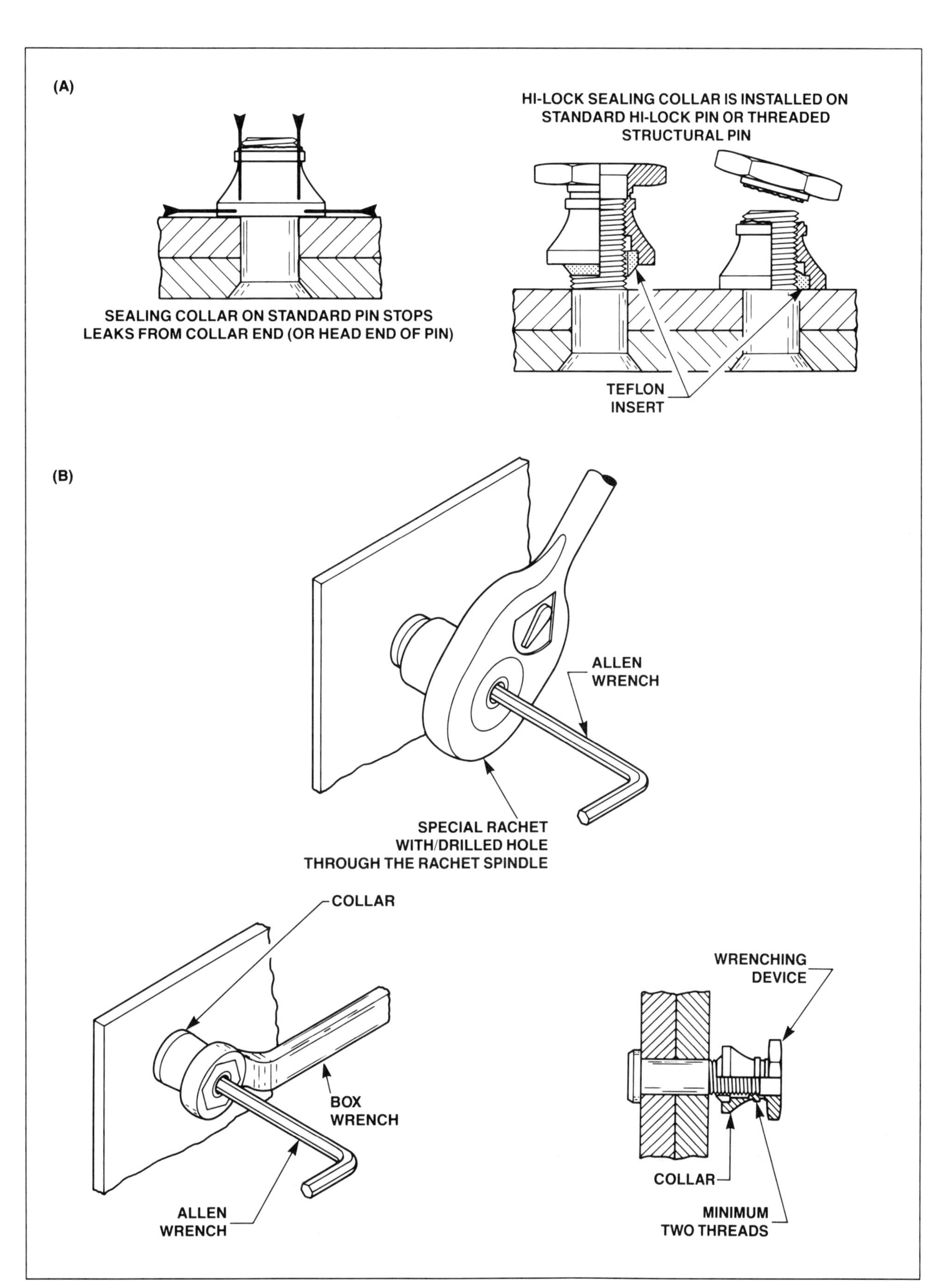

Fig. 5-6 Hi-Lok hand installation tools.

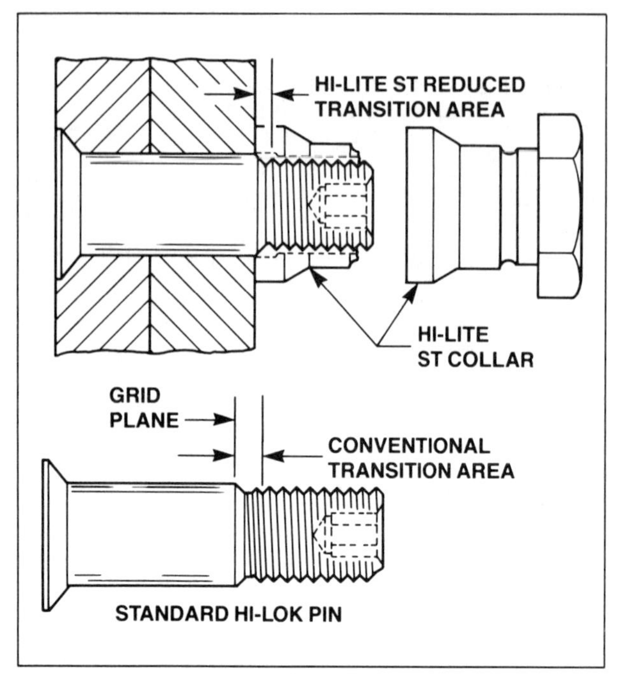

Fig. 5-7 The Hi-Shear Hi-Lite Fastener. Hi-lite is lighter weight than standard Hi-Lok pin. Hi-Lite ST collar shown above is only one of four available choices. Main difference between Hi-Lite and Hi-Lok is transition area from shoulder to first locking thread. This shorter area increases Hi-Lite's fatigue life.

The Hi-Lite is visually inspected by examining the place where the shear nut fractures away after installation. At least one groove on the stud should be threaded into the last groove on the collar.

E. Lockbolt Fasteners

The lockbolt is a fastener often used where high strength is required in the superstructure of the aircraft. Until the lockbolt was developed, special fasteners could not handle tension, that is, bearing loads.

There are two styles of lockbolts, the conventional and the blind. The conventional lockbolt style is available in pull and stump types. Both pull and stump types have two head styles, countersunk and flat. Conventional lockbolts are further classified as shearing and bearing types. When installed, the lockbolt is held into position by a crushable collar similar to the kind used with Hi-Shear rivets. The collar is made of material compatible with the stud, making these fasteners capable of carrying both bearing and shearing loads.

The shear-load lockbolt can be distinguished from the tension-type lockbolt by the number of grooves on the shank where it meets the collar. The shear-load lockbolt has two locking grooves located below the taper of the shank; the tension-type has four

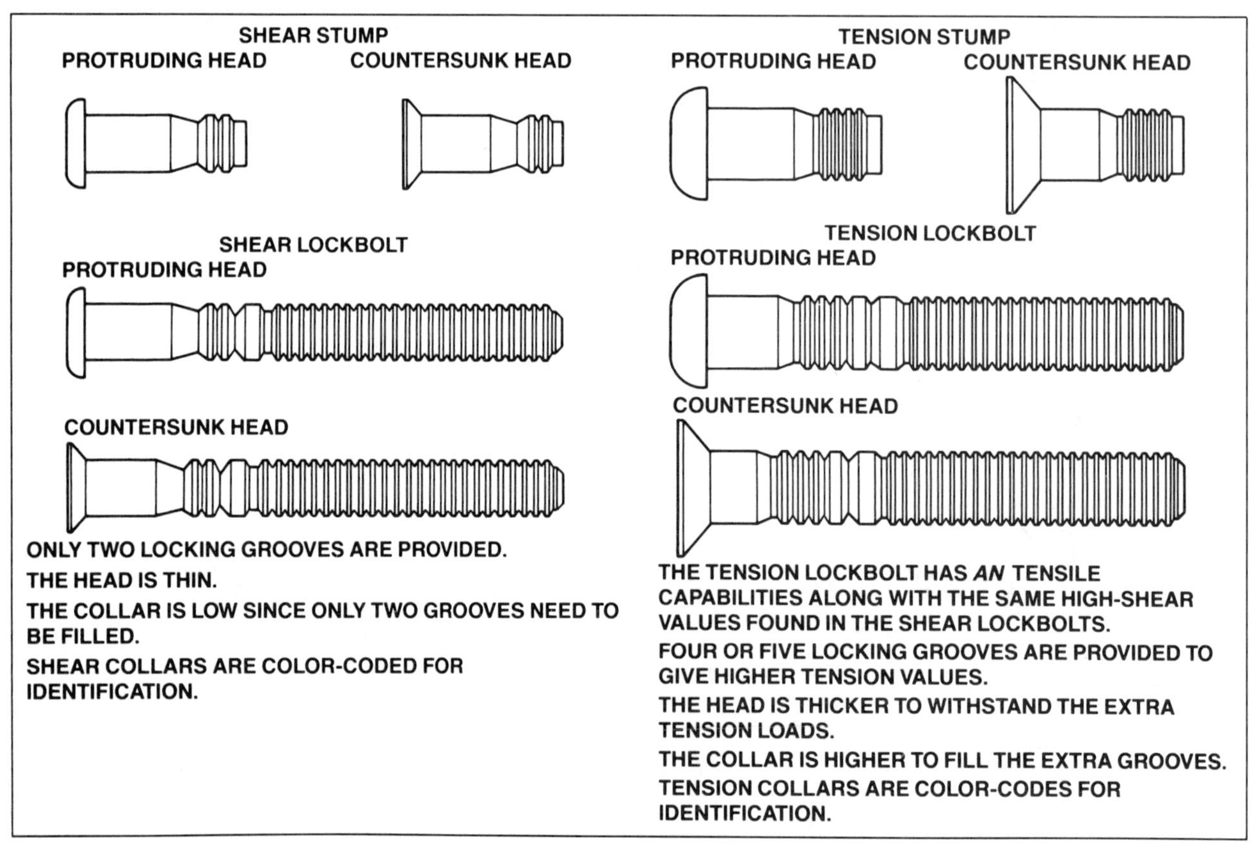

Fig. 5-8 Two general classifications of lockbolts and stumps for use in tension applications.

or five. The grooving on the tension-type lockbolt allows it to withstand high bearing loads. These two types of lockbolts are shown in Figure 5-8.

The stump type of lockbolt is the same as the pull type, except that the pulling groove section is snapped off, and the fastener is installed like the Hi-Shear rivet. The lockbolt is tapped into the recommended hole and a collar is slipped over the locking groove. Then a special gun set and large bucking bar are used to form the retaining collar (Figure 5-9).

Special tools are used for the installation of pull-type lockbolts. These tools must have the correct nose assembly installed that will correspond to the lockbolt being installed. Standard lockbolts and corresponding nose assemblies are shown in Figure 5-10. Proper installation also requires determination of correct grip length (Figure 5-11).

Like all special fasteners, the lockbolt requires special tooling and is designed for specific structural uses. Because lockbolt tooling is expensive, lockbolts are most likely to be used by larger aircraft repair stations and aircraft manufacturers. Regular AN bolts cannot be used as a substitute without the approval of the aircraft manufacturer.

1. Blind Lockbolts

Blind lockbolts have exceptional strength and sheet ***pull-together*** characteristics. They are used where only one side of the work is accessible and, generally, where it is difficult to drive a conventional rivet or fastener. Not to be used indiscriminately, they must be approved by an inspector whenever they are used on aircraft superstructures. The blind lockbolt comes completely assembled. It is installed in the same manner as the pull-type lockbolt.

F. Cherrybuck Titanium Shear Pins

Cherrybucks are lightweight, bimetallic, one-piece shear pins with a driven shear strength of 95 KSI. The head and shank are made of 6AL-4V titanium alloy. The upper shank must have an interference fit.

The tail portion is made of Ti/Col (titanium-columbium), which is more ductile to allow formation of the shop head. The combination of these two alloys saves weight and eliminates the need for a locking collar, because the lower end of the shank can be driven like a solid-shank rivet (Figure 5-12).

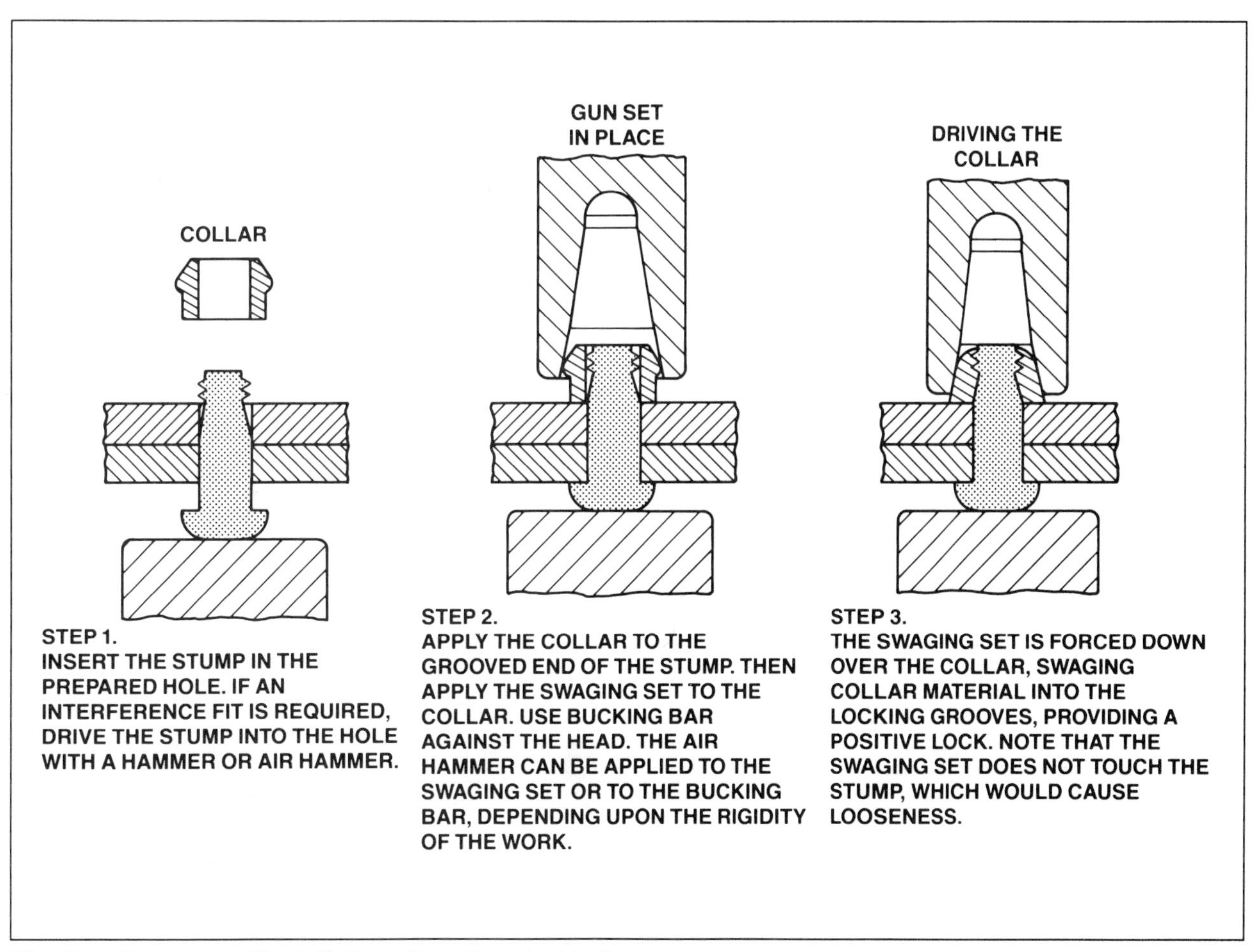

Fig. 5-9 Sequence of installation, stump type lockbolt.

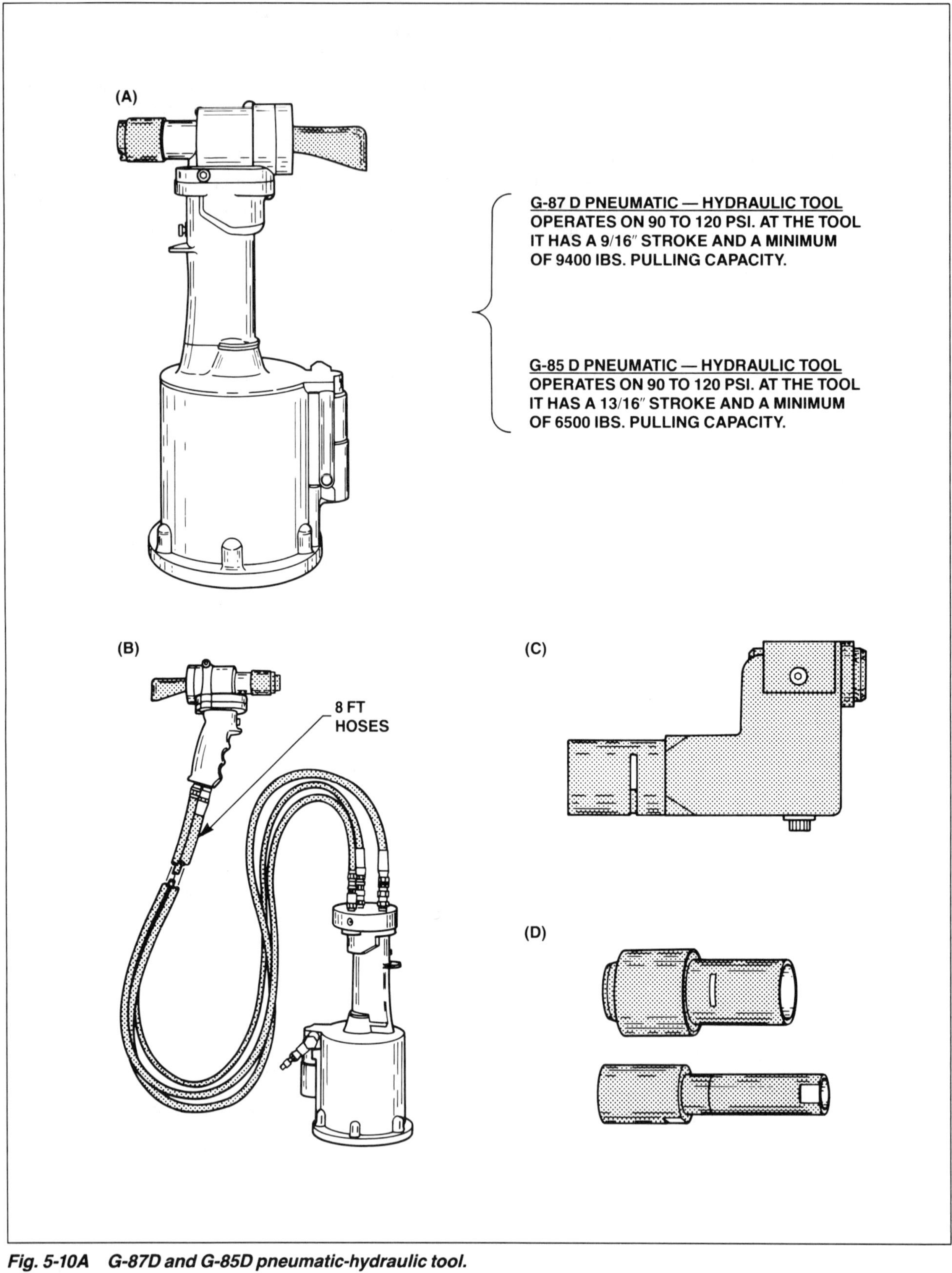

Fig. 5-10A ***G-87D and G-85D pneumatic-hydraulic tool.***
Fig. 5-10B ***G-85D-S pneumatic-hydraulic tool with 8 ft. hoses.***
Fig. 5-10C ***Nose assembly.***
Fig. 5-10D ***Adapters.***

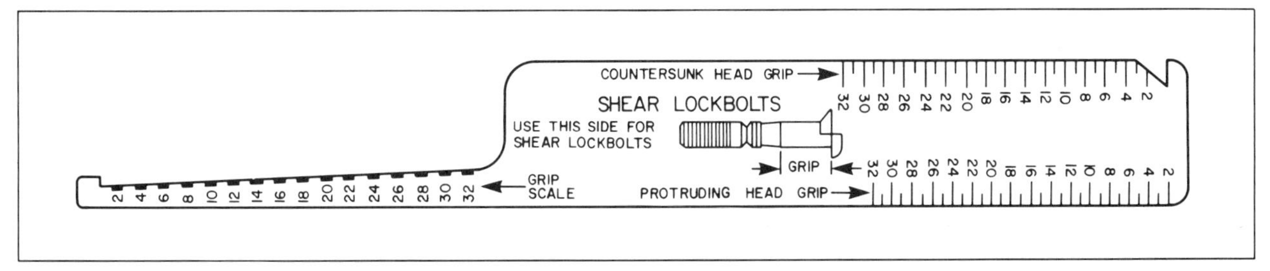

Fig. 5-11 A simple, self-explanatory gage for determining material thickness and proper lockbolt grip length.

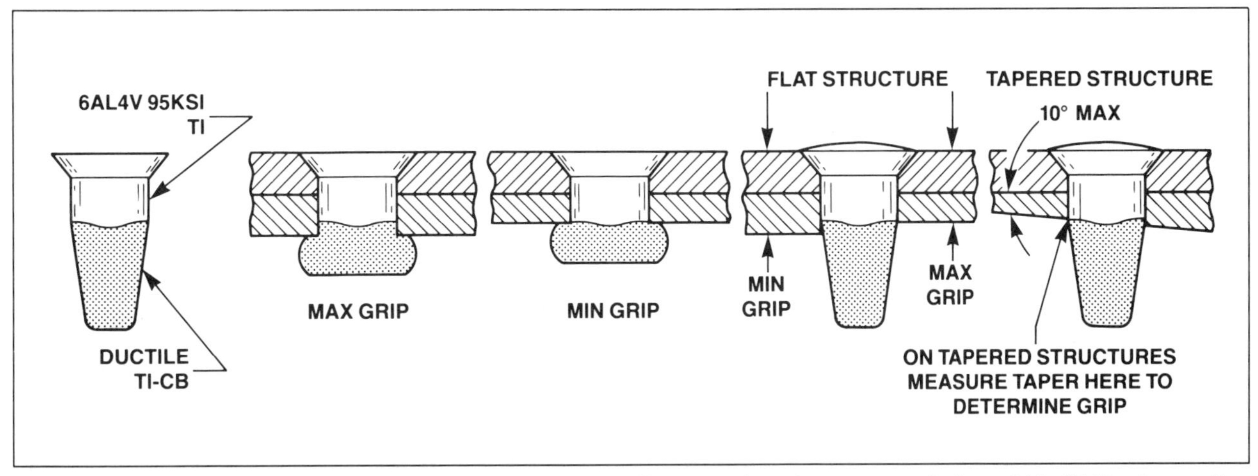

Fig. 5-12 Cherrybuck fasteners. Cherrybuck is a bimetallic, one piece fastener made from 6AL-4V-titaninum alloy with 95 KSI shear strength and a ductile TI-CB titanium-columbium lower shank shop head section. The upper shank must be fitted as an interference fit. Because the bucktail is driven, a Cherrybuck can be installed on inclined surfaces of up to 10°.

The design of the Cherrybuck allows it to be installed on an inclined surface up to 10° without a special locking collar to secure it. There is no danger of foreign object damage (FOD) occurring from a Cherrybuck because a conventional bucktail is formed on the end of its shank. Some ***two-piece fasteners*** have been known to shed their collars inside of a jet engine nacelle.

The Cherrybuck is available in five diameters, ranging from 5/32 to 3/8-inch in increments of 1/32-inch. The length of Cherrybucks is measured in increments of 1/16-inch. The Cherrybuck is available in three head styles: MS20426 flush, NAS1097 flush, and protruding. They are lubricated with chlorine-free cetyl alcohol for ease of fitting into holes.

The Cherrybuck is used structurally on some corporate aircraft. One of the main advantages of using the Cherrybuck in small aircraft shops is that it can be installed, ensuring an interference fit, with a conventional solid-shank rivet gun and bucking bar.

G. Huckbolt Fasteners

The ***Huckbolt*** can handle full tension as well as shear loads. It has a tensile strength up to 108 KSI. Maximum shear strength is achieved by installing the lockbolt in a close-tolerance hole.

The Huckbolt is available in two styles: a pull and a stump type (Figure 5-13). The difference between the two is that the stump-type shank does not have a pulling section. The pull type is installed with a special installation tool which grips the pulling stem while a tapered foot presses the locking collar into the Huckbolt grooves. When in place, the pulling stem snaps off.

Inspection consists of a visual examination of the swaged locking collar and the fit of the MFG head. The heads must be flat and binding against the surfaces they are clamping together. The stump type is installed by squeezing or driving with a rivet gun.

The Huckbolt has a wide variety of head styles, the flat and flush heads being the most commonly used. The Huckbolt is made from two widely used alloys: A-286 corrosion-resistant steel and 6AL-4V titanium. Huckbolts are available in diameters ranging from 5/32 to 3/8-inch, in increments of 1/32-inch. Their lengths are measured in increments of 1/16-inch, and they can be used on thicknesses ranging from .020 to 2 inches. The Huckbolt collar is made from the aluminum alloys 2219 and 2024, stainless steel A-286 and Monel steel.

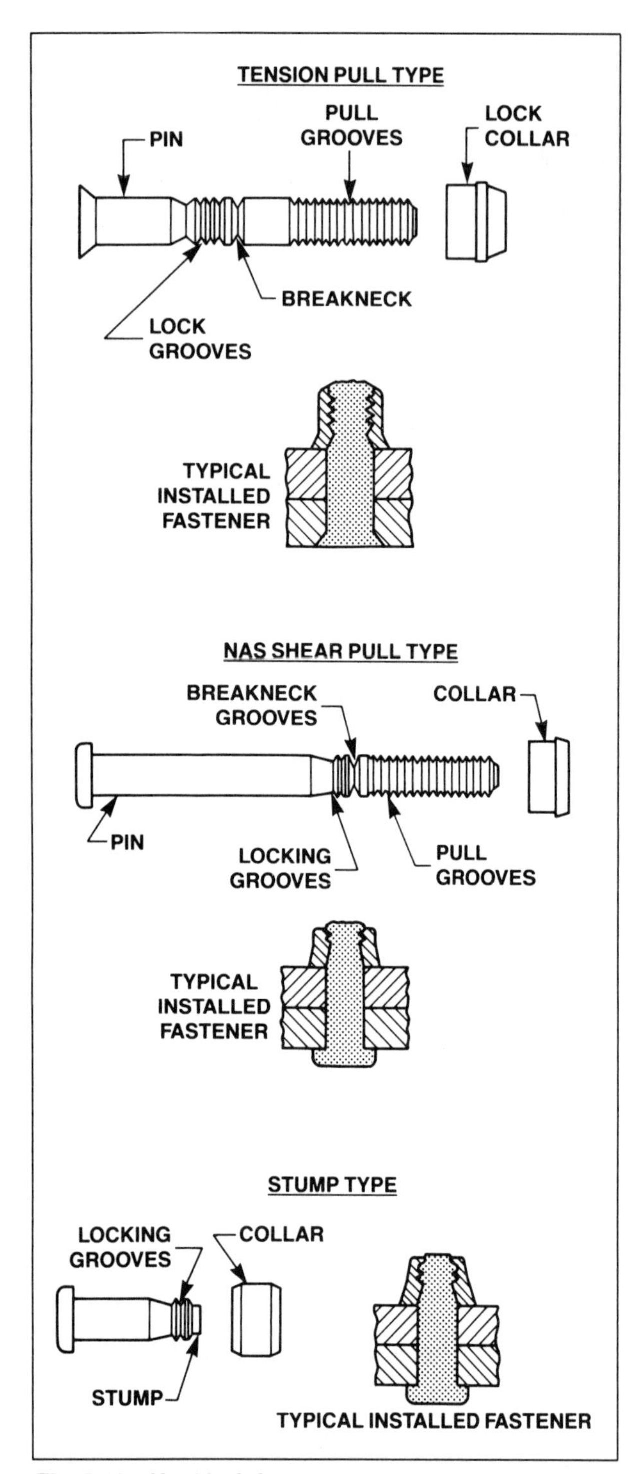

Fig. 5-13 Huckbolt fasteners.

H. Taper-Loks

The strongest special fastener, used in the construction of many standard and wide-body jet aircraft, is the Taper-Lok, a breakthrough in jet age fasteners. The Taper-Lok was developed to improve the structural fatigue life of modern jet aircraft. These fasteners differ from conventional fastening systems in that, when they bind structures together, they greatly add to the fatigue life of the aircraft. Thus, the Taper-Lok is used where the ultimate in strength is needed.

For example, in the center section of the Boeing 747, 45,000 Taper-Loks are installed in its body fittings, keel beam, ***fairing*** beam chords, and three main spars. The Taper-Lok is effective in absorbing either static or dynamic loads. The whole installation area is preloaded to the point where, under normal overload conditions, the joint will not indicate any significant strain because, being in a prestressed condition, the area never relaxes.

The Taper-Lok was designed to copy the action of an installed solid-shank rivet and an interference fit, ***straight-pin fastener.*** It has a tapered, cone-shaped shank, which is inserted into a precision-drilled, tapered hole. The shank is forced to slide down the tapered hole, like a cork in a bottle, as the washer nut draws tight. This action provides uniformly controlled interference preloading. Figure 5-14 shows how static preload, radial ***compression,*** and peripheral tension result from the interference fit of the Taper-Lok shank.

The lines of force are generated by the clamp-up action between the MFG head and the washer nut. The lines of force are amplified due to the taper of the hole and the tightening of the fastener. As a result of this action, in the conical hole, radial compression occurs. This action causes the area to become pre-stressed.

The proper installation of the Taper-Lok (Figure 5-14) provides excellent clamp-up action between its head and washer nut. The holes for the Taper-Lok are prepared with a tapered drill and tapered 1/4-inch per foot. This small amount of taper is enough to cause a preload to occur. The installation is finished off by tapping the Taper-Lok into the hole and tightening it into position with a washer nut.

Taper-Loks are available in two head styles — flush heads and protruding heads, which are used for shear and tension loads. The shank diameters are measured in increments of 1/16-inch. The diameters are available in a range from 3/16 to 1/2-inch. The lengths of the Taper-Lok have a range associated with their diameters. For example, the 3/16-inch diameter fastener is available in lengths of six-teenths of an inch. Taper-Loks are made from steel, A-286 stainless steel, titanium alloy, and nickel base alloys.

There are two types of Taper-Loks. The threaded type uses a washer nut and the non-threaded type uses a swaged collar. There are six different types of washer nuts available for Taper-Lok installation. All are strong, lightweight unit assemblies

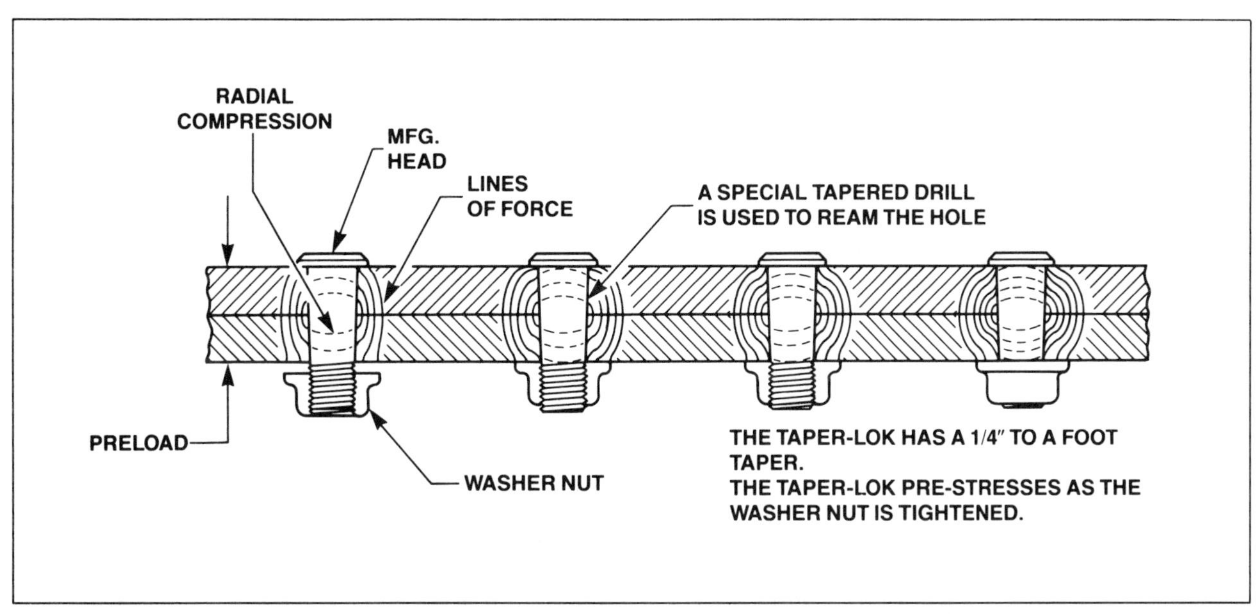

Fig. 5-14 Static pre-loading and installation of a Taper-Lok. Lines of force are generated by clamp-up action between MFG head and washer nut. Lines of force are amplified due to taper of the hole and the tightening of fastener. As a result of this action, in the conical hole, radial compression occurs. This action causes the area to become pre-stressed.

in which the nut can turn freely. The washer nuts are made from different alloys so as to be compatible with the Taper-Lok.

I. Blind Fasteners

Rebuilding aircraft structures often requires the removal of damaged sections of the airframe, wing, or tail group. When the new or rebuilt parts are re-riveted, solid-shank rivets should be used to replace solid-shank rivets wherever possible.

However, in areas which are inaccessible from one side, blind fasteners can be used. Many different types of blind fasteners are available for various situations. Grip length gauges should be used to determine the length of the fastener to be installed. On critical areas, the fasteners can be used only when specified on the drawing or with the manufacturer's written approval.

Blind fasteners require special installation tools and special handling. Because all blind rivets are lubricated with cetyl alcohol when shipped, they must be kept in their original containers until used. If taken from the container, they may lose lubrication or become coated with dust or waste material. This would make them difficult to install and could cause the stem to break before the rivet is fully pulled up.

1. Pop Rivets

The Pop rivet is used as a temporary fastener when a tight, well drawn-up rivet joint is needed. Pop rivets are made out of aluminum, plain steel, stainless steel, copper, and Monel steel. The Pop rivet assembly consists of a rivet and a pulling stem that resembles a finishing nail. The Pop rivet is available in three different diameters: 1/8, 5/32, and 3/16-inch. The length designations are usually short, medium and long. Pop rivets have two head styles: flat and unisink (Figure 5-15).

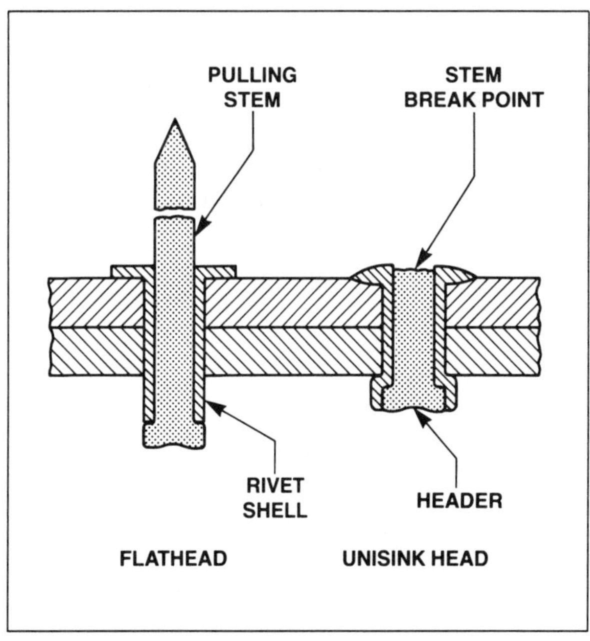

Fig. 5-15 Pop rivet. Standard sizes: Diameter 1/8, 5/32 and 3/16. Lengths vary acording to short, medium and long. Head styles are flat and unisink. Materials used are aluminum, steel, Monel and copper.

2. Rivnuts

The ***Rivnut,*** manufactured by the BF Goodrich Company, was first used on the deicer boots of early models of the DC-3. It served as an anchor nut for the machine screws used to hold down the deicer boots. It was also used to install wing fairings and access plate covers. The tool used for installation is a drawbolt equipped with a set of threads that, when squeezed, will form a ***bulbed end*** which serves as the anchor nut. Figure 5-16 shows a typical Rivnut installation. The Rivnut is not used extensively on modern aircraft; more versatile hardware, such as the ***plate nut,*** is preferred.

3. Friction Cherrylock Rivets

The friction Cherrylock rivet is an older special blind fastener which is still used by many sheet metal mechanics. Friction is the only thing that holds the stem of this fastener in place. In locations where vibrations occur, the stem can work loose and fall out. For this reason, whenever the Cherrylock friction rivet is used to replace a solid-shank rivet, it must be 1/32-inch larger in diameter. The friction Cherrylock rivet is available in three diameters: 1/8, 5/32, and 3/16-inch.

The friction Cherrylock rivet is constructed of two parts: The rivet, with either a universal or a countersunk head, and a ***mandrel*** or pulling stem that runs through the center of the hollow rivet for installation purposes. The pulling stem is designed with an expanding section which causes the shank to fill the hole when it is being drawn up. On the bottom of the puller is a part called the header, which forms the bucktail on the blind side. The upper portion of the pulling stem has a break point which lets go when the rivet is pulled to its maximum. When the break point snaps, the remaining part of the stem protrudes up from the manufacturer's head and has to be clipped and filed down flush.

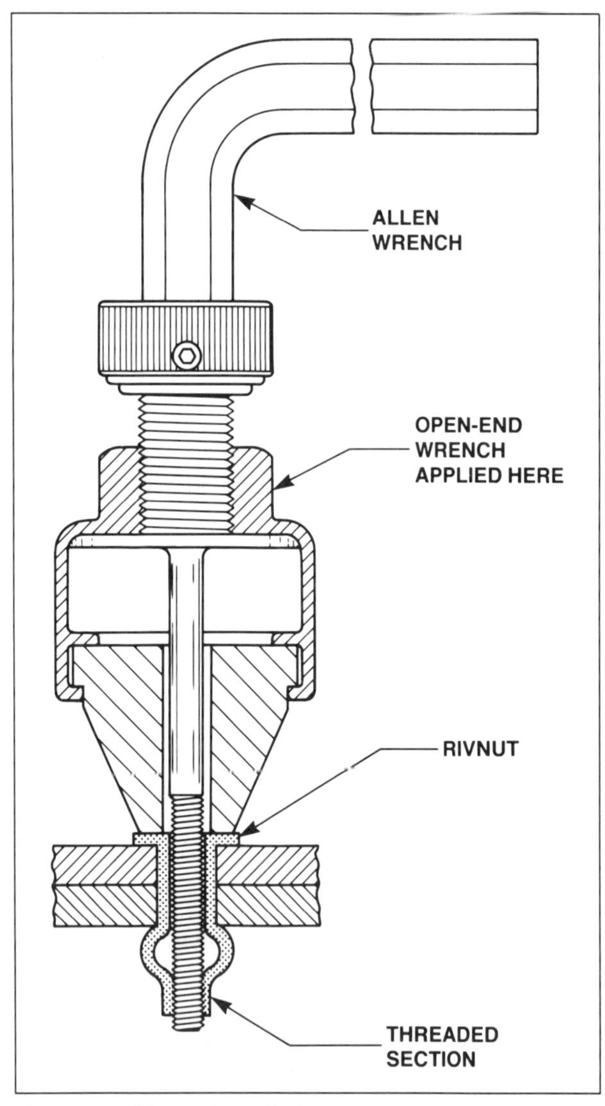

Fig. 5-16 Rivnut installation.

J. Mechanical Lock Blind Rivets

1. Bulbed Cherrylock Rivets

The ***bulbed Cherrylock rivet*** is an improvement over the friction Cherrylock rivet because its stem is locked into place with a retaining ring. The bulbed Cherrylock, with its ***locked spindle,*** has many uses, such as ***thin-sheet,*** skin-to-spar stack-up, and wing leading edge applications. Its large blind head, high strength, and locked spindle make the bulbed Cherrylock structurally sound enough to replace solid-shank rivets. Figure 5-17 shows an installation of the bulbed Cherrylock fastener.

The bulbed Cherrylock rivet is available in two head styles: countersunk and universal. Its diameters are measured in increments of 1/32-inch. It is available in three standard diameters of 1/8, 5/32 and 3/16-inch, and an oversize 1/64-inch diameter. Bulbed Cherrylock rivet lengths are measured in increments of 1/16-inch. The rivet portion of the bulbed Cherrylock rivet is made of 2017 or 5056 aluminum alloy, Monel, or stainless steel.

The main disadvantage of the Cherrylock mechanical rivet is that a special tooling head is required to install different sizes and head styles. The Cherrymax rivet, the latest Cherry blind fastener, can be installed with one tool.

2. Cherrymax Rivets

The Cherrymax fastener is a reliable, high-strength, structural rivet. It is inexpensive, easy to install and to inspect, and a suitable replacement for solid-shank rivets.

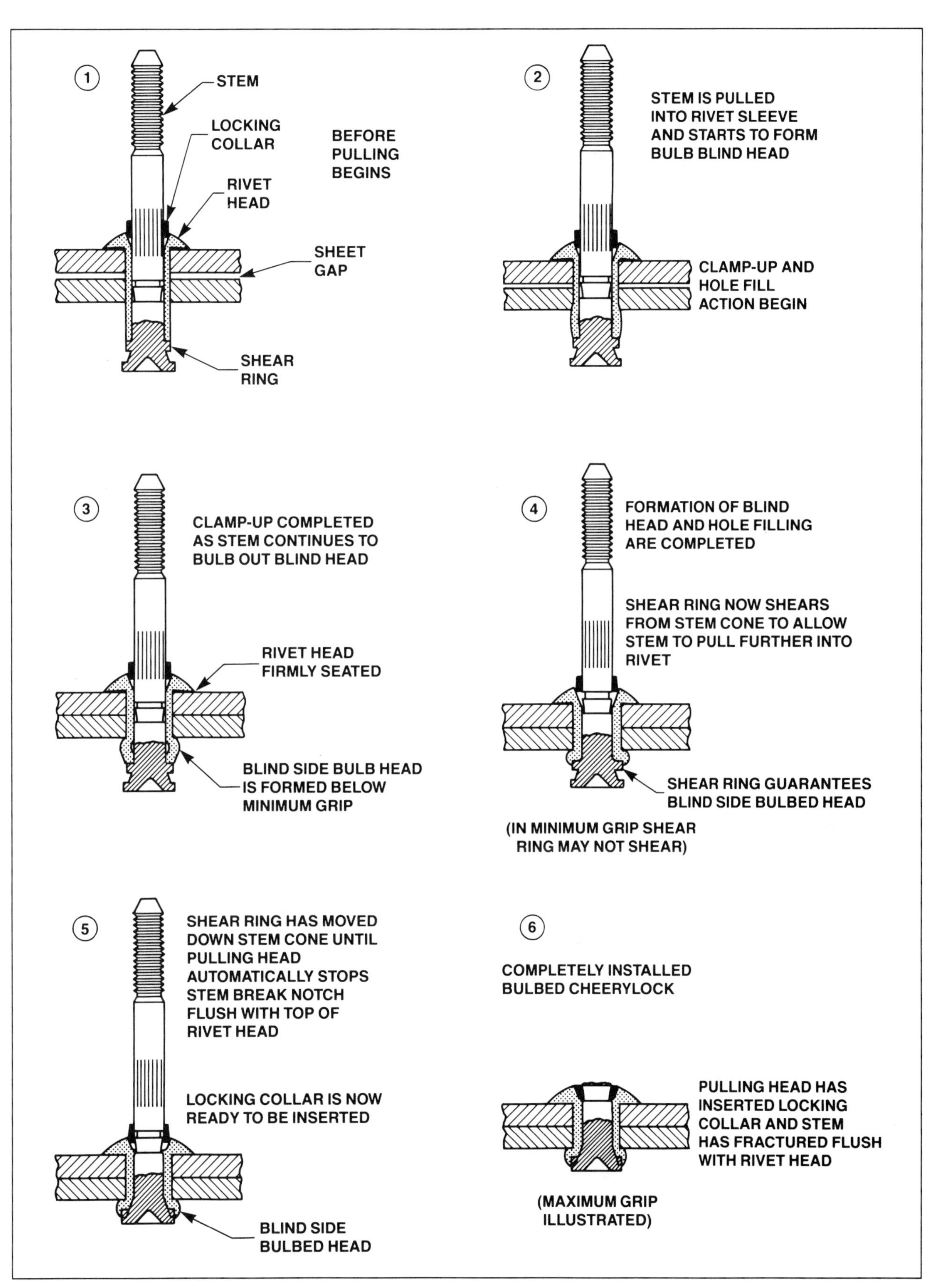

Fig. 5-17 Bulbed Cherrylock rivet.

Bulbed CherryMAX Rivets are available in nominal and oversize 1/8, 5/32, 3/16, and 1/4″ diameters in the material combinations shown below:

MATERIALS		ULTIMATE SHEAR STRENGTH	MAXIMUM TEMPERATURE
SLEEVE	STEM		
5056 ALUMINUM	ALLOY STEEL	50,000 PSI	250°F
5056 ALUMINUM	CRES	50,000 PSI	250°F
MONEL	CRES	55,000 PSI	900°F
INCO 600	INCO X-750	75,000 PSI	1400°F

CherryMAX, with its light-weight, simplified tooling, replaces solid aluminum rivets.
Comparative shear allowables, CherryMAX vs. solids, are shown below:
Ultimate Static Joint Strength Loads (lbs.) per MIL-HDBK-5 Criteria

RIVET SHANK DIAM.	RIVET PART NUMBER	UNIVERSAL & 100° FLUSH HEAD ALUMINUM RIVETS IN 2024 T3 ALCLAD ALUMINUM. 100° FLUSH HEAD RIVETS IN MACHINE COUNTERSUNK SHEET.														
		SHEET THICKNESS														
		.016	.020	.025	.032	.040	.050	.063	.071	.080	.090	.100	.125	.160	.190	.250
1/8″ SOLIDS	MS 20426 AD FLUSH	—	163	221	272	309	340	363	373	388	388	388	388	388	388	388
	MS 20470 AD UNIV.	—	—	357	374	386	388	388	388	388	388	388	388	388	388	388
1/8″ NOM.	CR 3212 FLUSH M7885/3-4	—	—	—	254	299	355	428	460	491	526	561	644	664	664	664
	CR 3213 UNIV. M7885/2-4	152	206	272	342	392	454	518	538	561	586	611	664	664	664	664
1/8″ O.S.	CR 3242 FLUSH M7885/7-4	—	—	218	285	363	443	537	594	620	642	665	722	801	814	814
	CR 3243 UNIV. M7885/6-4	187	245	319	395	451	520	610	641	667	697	726	801	814	814	814
5/32″ SOLIDS	MS 20426 AD FLUSH	—	—	250	348	418	479	523	542	560	575	596	596	596	596	596
	MS20470 AD UNIV.	—	—	—	551	575	593	596	596	596	596	596	596	596	596	596
5/32″ NOM.	CR 3212 FLUSH M7885/3-5	—	—	—	—	395	465	556	612	676	721	764	873	1005	1030	1030
	CR 3213 UNIV. M7885/2-5	—	236	320	439	532	610	711	772	810	841	872	951	1030	1030	1030
5/32″ O.S.	CR3242 FLUSH M7885/7-5	—	—	—	345	442	561	684	759	838	922	950	1020	1120	1205	1245
	CR 3243 UNIV. M7885/6-5	—	288	378	506	608	694	804	873	949	987	1025	1115	1245	1245	1245
3/16″ SOLIDS	MS 20426 AD FLUSH	—	—	—	—	525	628	705	739	769	795	818	853	862	862	862
	MS 204707 AD UNIV.	—	—	—	—	804	836	862	862	862	862	862	862	862	862	862
3/16″ NOM.	CR 3212 FLUSH M7885/3-6	—	—	—	—	—	584	693	760	835	920	994	1125	1305	1435	1480
	CR 3213 UNIV. M7885/2-6	—	—	362	502	661	784	904	978	1060	1140	1175	1270	1400	1480	1480
3/16″ O.S.	CR 3242 FLUSH M7885/7-6	—	—	—	—	505	647	823	906	998	1105	1205	1315	1425	1525	1685
	CR 3243 UNIV. M7885/6-6	—	—	429	576	741	858	985	1065	1155	1255	1315	1420	1570	1685	1685
1/4″ SOLIDS	MS 20426 AD FLUSH	NOT LISTED ON MIL-HDBK-5 TABLES														
	MS 20470 AD UNIV.	—	—	—	—	—	1430	1495	1520	1545	1555	1555	1555	1555	1555	1555
1/4″ NOM.	CRL 3212 FLUSH M7885/3-8	—	—	—	—	—	—	—	1080	1185	1295	1410	1685	1945	2150	2535
	CR 3213 UNIV. M7885/2-8	—	—	—	606	819	1085	1340	1440	1550	1675	1800	2040	2215	2365	2615
1/4″ O.S.	CR3242 FLUSH M7885/7-8	NOT LISTED ON TABLES — CONTACT TECH FOR VALUES														2925
	CR 3243 UNIV. M7885/6-8	NOT LISTED ON TABLES — CONTACT TECH FOR VALUES														2925

ALLOWABLE DATA SOURCES: MS 20426 — MIL-HDK-5 TABLE 1.1.2.2(d)
MS 20470 — MIL-HDBK-5 PARAGRAPH 8.1.2.1
CR 3212(M7885/3) — MIL-HDBK-5 TABLE 8.1.3.2.2(q)
CR 3213(M7885/2) — MIL-HDBK-5 TABLE 8.1.3.1.2(l)
CR 3242(M7885/7) — MIL-HDBK-5 TABLE 8.1.3.2.2(p)
CR 3243(M7885/6) — MIL-HDBK-5 TABLE 8.1.3.1.2(k)

Fig. 5-18 Comparison of shear strengths of Cherrymax rivets and shear strengths of equivalent solid-shank rivets.

In Figure 5-18, the shear strengths of Cherrymax rivets are compared to the shear strengths of the equivalent solid-shank rivets.

The Cherrymax has five parts: A serrated pulling stem with a breakaway notch, a driving anvil to insure a visible mechanical lock, a locking collar with an internal grip adjustment cone, a rivet sleeve with a recessed head to receive the locking ring, and a bulbed blind head (Figure 5-19). The driving anvil, which looks like a washer, eliminates wear and tear on the installation tool. This allows the use of one pulling head on all diameters.

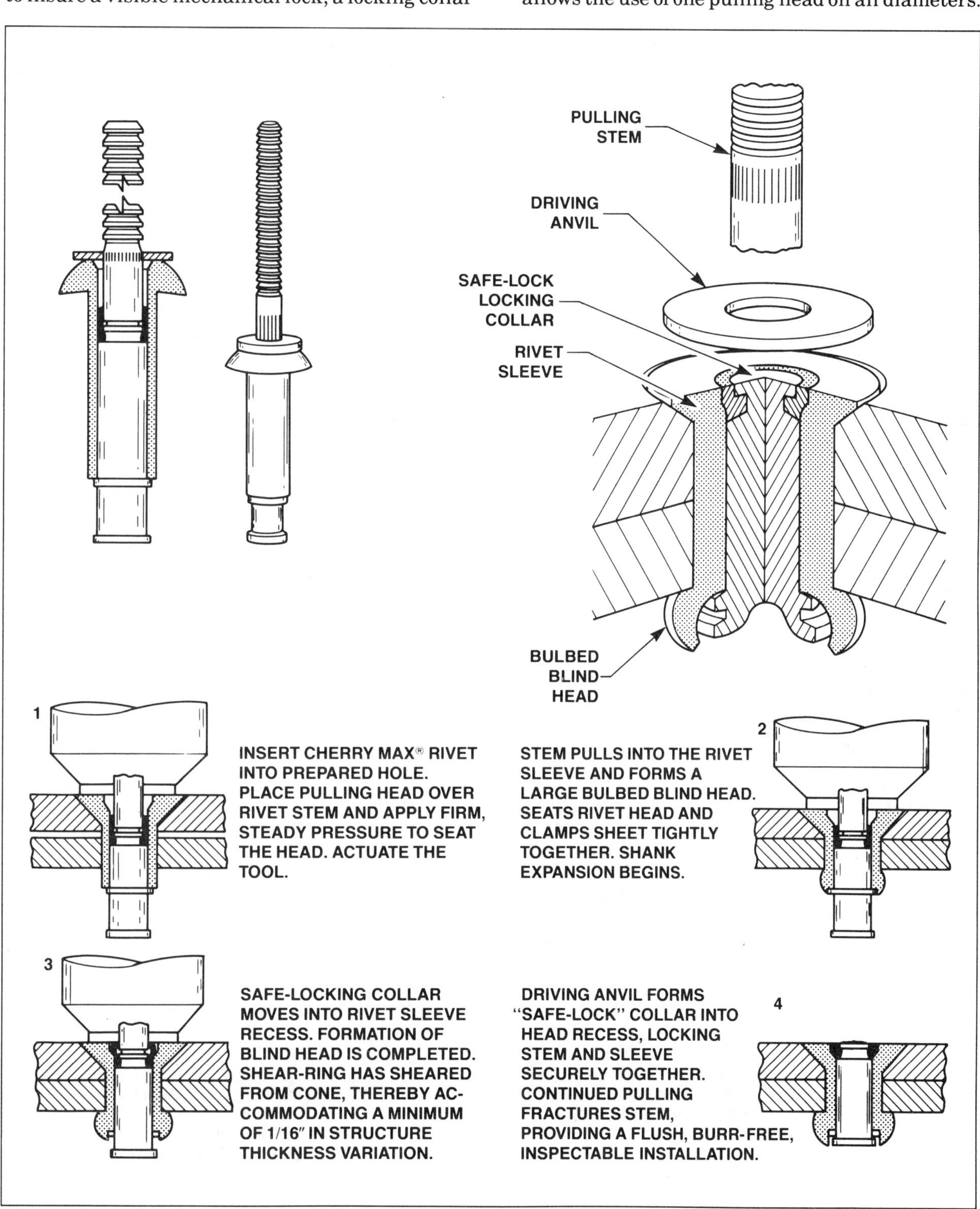

Fig. 5-19 Cherrymax.

The Cherrymax rivet is available in six diameters: 1/8-inch nominal and oversize, 5/32-inch nominal and oversize, and 3/16-inch nominal and oversize. It is manufactured in five head styles: Universal MS20470, 100° flush MS20426, 100° flush NAS1097, unisink, and 120° flush. Figure 5-20 shows a typical part number breakdown and the various head styles and uses of the Cherrymax rivet.

K. Olympic-Lok® Fasteners

The Olympic-Lok fastener is a lightweight, mechanically locked, spindle-type blind rivet. Olympic-Loks are equipped with a locking ring, integral to the pulling stem, which is used to lock the stem into position. It also provides a large formed head on the blind side, which improves the

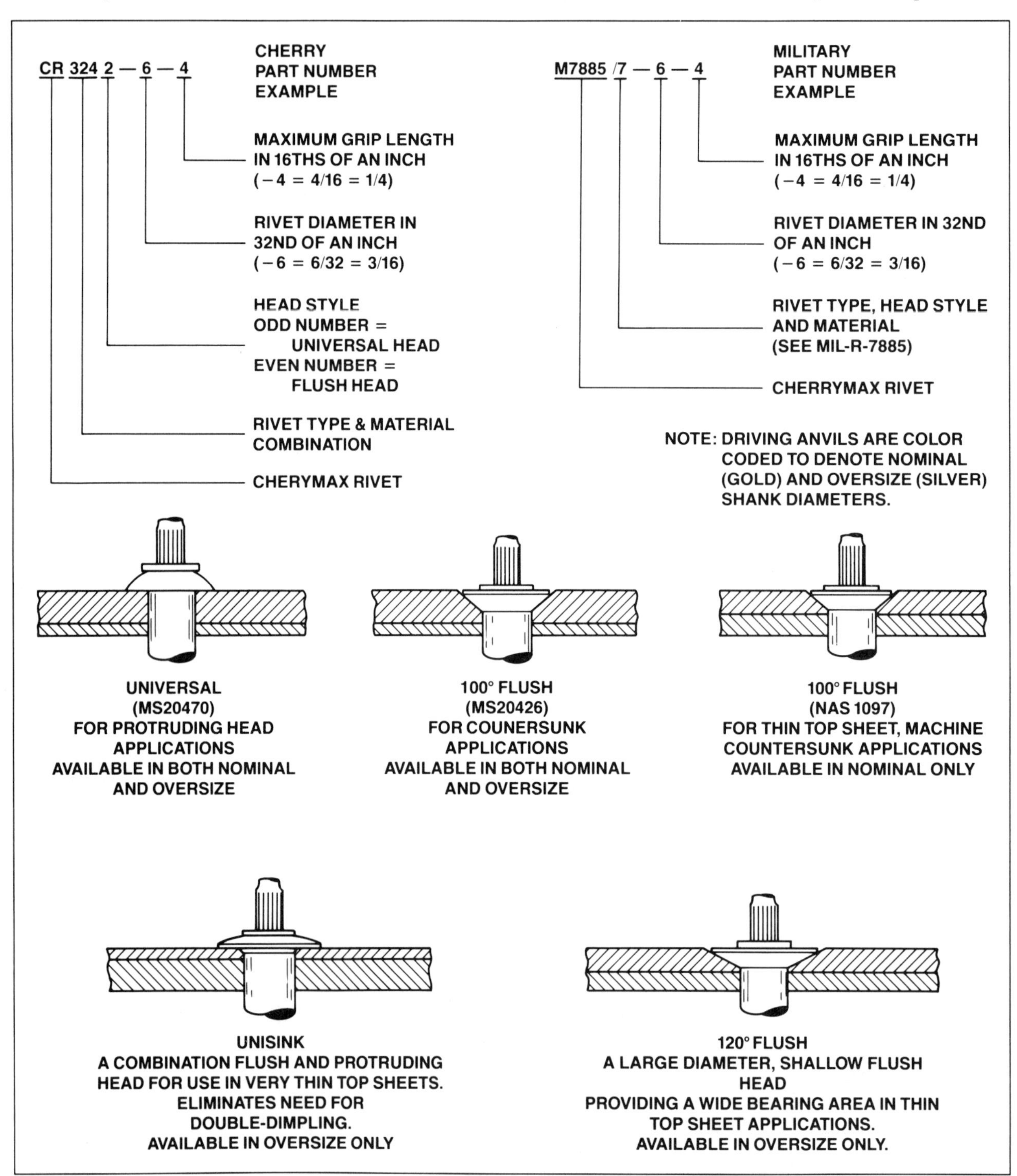

Fig. 5-20 Part number breakdown and various head styles and uses of Cherrymax rivet.

bearing strength of the joint. The clamp-up action and shank expansion improve shear strength.

The Olympic-Lok blind fastener is a total fastening system which includes fastener and tools. The Olympic-Lok pulling heads are designed to pull protruding and countersunk heads in the same diameter range. The tool is designed to permit the installation of fasteners from one side of open or limited-access structures. Figure 5-21 shows the installation of an Olympic-Lok blind rivet with the locked spindle.

The Olympic-Lok blind fastener is available in three head styles: Universal protruding, 100° flush countersink, and 100° flush shear, and three diameters: 1/8, 5/32 and 3/16-inch. The three diameters are available in eight different alloy combinations of 2017-T4, A-286, 5056, and Monel. The Olympic-Lok locked spindles are made from the same materials that the sleeves are made.

L. Huck Fasteners

There are two kinds of Huck blind fasteners: Huck NAS1900 blind rivets and Huck-Clinch blind rivets. The locking collar of the NAS1900 is connected directly to the rivet head. This locking collar is swaged into a conical hole around the rivet head, simultaneously filling a locking groove around its spindle. Figure 5-22 shows the installation of the NAS1900 Huck rivet.

The NAS 1900 Huck rivet is available in two head styles: protruding and flush. It is available in four diameters: 1/8, 5/32, 3/16 and 1/4-inch. The diameters are measured in increments of 1/32-inch and lengths are measured in 1/16-inch increments. They are manufactured in three different combinations of alloys: a 5056 aluminum sleeve with a 2024 aluminum alloy pin, an A-286 corrosion-resistant steel sleeve with an A-286 pin, and a Monel 400 sleeve with an A-286 pin.

The Huck-Clinch fastener has the ability to draw up the metal to a tight and binding joint. The locking of the spindle into position is accomplished by the action of a stop anvil pressing down on the rivet head. The pressing action of the stop anvil and the draw-up of the spindle cause the rivet to expand, thus filling the locking area around the spindle lock (Figure 5-23).

M. Summary Of Blind Fasteners

Each blind fastener is unique in construction. The Cherrymax, Olympic-Lok, Huck, and Cherry mechanical lock rivets all have stems that snap away when installed, leaving a locking device which keeps the stem from falling out, no matter what the situation or amount of vibration. The stems of these rivets must never be filed smooth because of the danger of removing the locking device. Always follow the manufacturer's recommendations regarding the handling of any blind fastener.

To remove a blind rivet, first file a flat spot on the stem section of the locking ring and center punch it. Then drill deep enough to remove the remaining lock ring. Use a small pin punch to tap out the stem. Remove the manufactured head by drilling deep enough to cut it free of the sleeve, and lightly tap out the remaining shank.

N. Special Fasteners: Access Panels And Cowlings

The cowling surrounding both jet and piston engines aids in cooling and streamlines the airflow around the engine. Due to the pressure generated by the air flow around an engine, its cowling must be locked on securely. Dzus buttons, Camlocs, and Airlocs are used to secure engine cowlings and access covers to various other parts of an aircraft.

1. Dzus Fasteners

Three main parts make up the ***Dzus fastener:*** The stud, spring, and grommet. The stud is made of steel and is available in three head styles: flush, oval, and wing. The spring is made of cadmium-plated, piano-hinge steel wire. The cadmium plating prevents dissimilar metal corrosion. The grommet is made from 1100 aluminum.

Dzus fasteners are manufactured in body diameters ranging from 3/16 to 3/4-inch (in increments of 1/16-inch) and in shank lengths ranging from .200 to 1 inch (measured in increments of one-hundredths of an inch).

On top of the manufactured head are the name "Dzus", a letter which denotes whether the fastener head is flush (F) or protruding (P), the diameter in sixteenths of an inch, and the length in hundredths of an inch. Figure 5-24 shows the parts and installation of a Dzus fastener.

Either a wing or slotted head Dzus fastener is used when attaching cowling or closing up access plates. The wing type is secured by turning it manually. The slotted Dzus button requires a common blade screwdriver. To secure a Dzus fastener, the curved slot on the end of its shank is made to straddle the lock spring prior to turning into the locked position. Secure lock-up is indicated by a clicking sound.

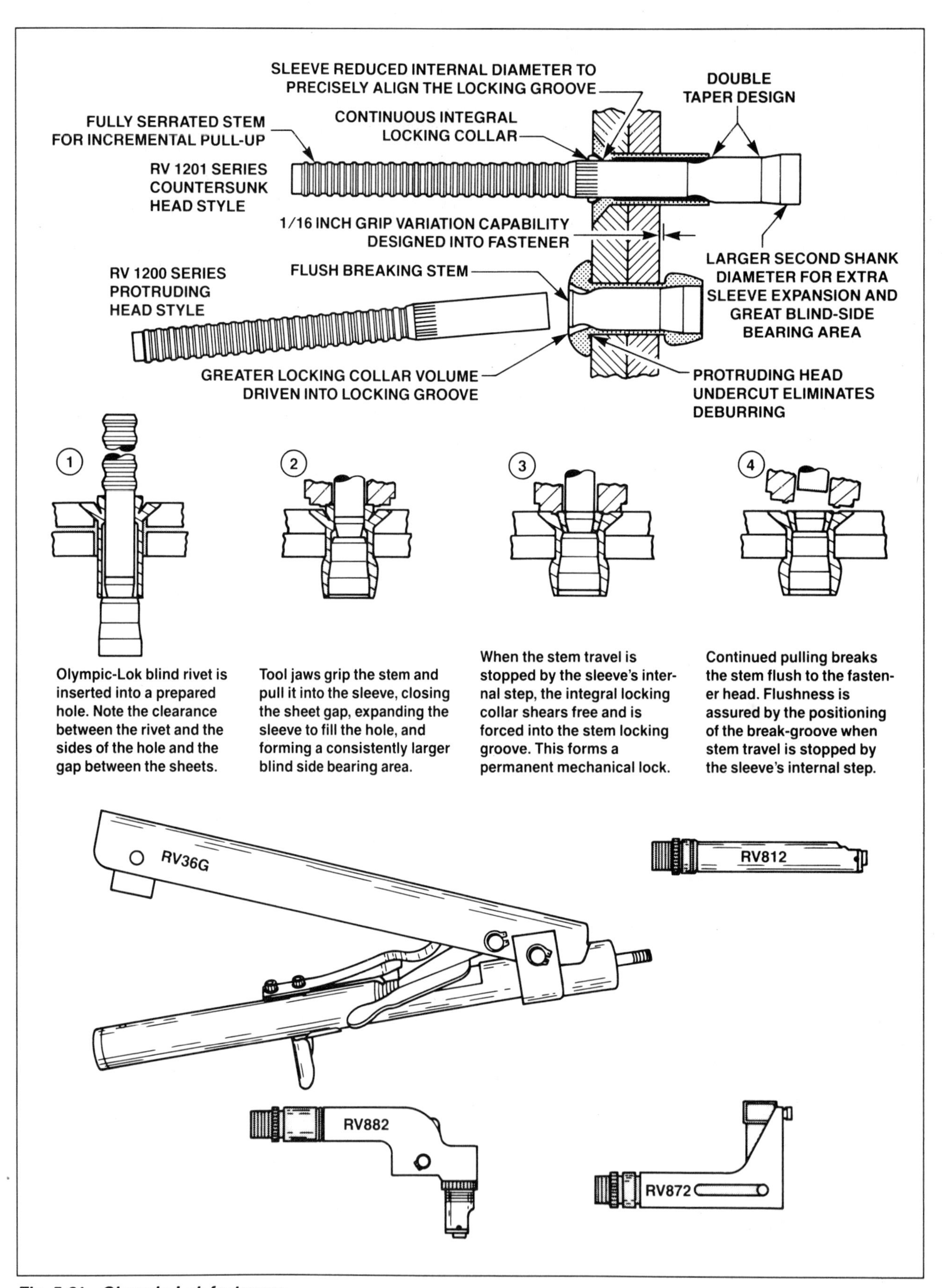

Fig. 5-21 Olympic-Lok fasteners.

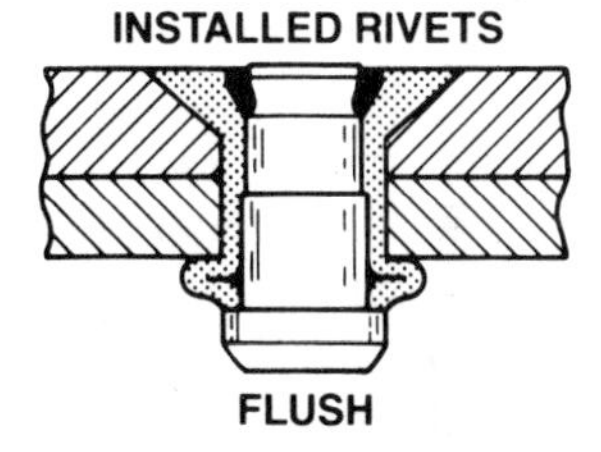

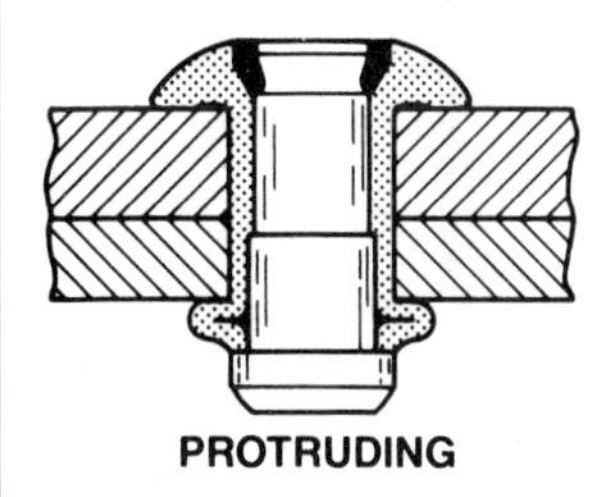

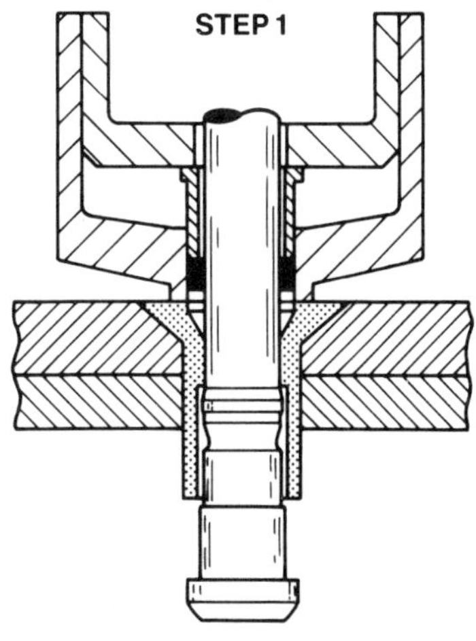

STEP 1. INSTALLATION SEQUENCE• FASTENER IS INSERTED INTO THE PREPARED HOLE. THE INSTALLATION TOOL IS APPLIED OVER THE FASTENER WHICH AUTOMATICALLY ENGAGES THE JAWS IN THE NOSE ASSEMBLY WITH THE PULL SERRATIONS OF THE SPINDLE. THE TOOL TRIGGER IS ACTUATED.

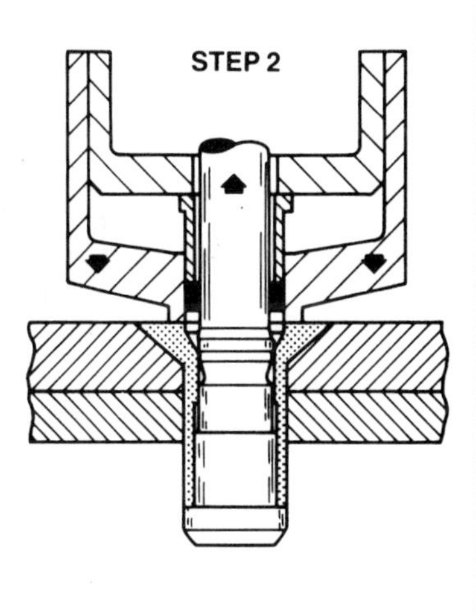

STEP 2. SLEEVE EXPANSION• THE INSTALLATION TOOL PULLS THE BALL LAND ON THE SPINDLE THROUGH THE SLEEVE. THE SLEEVE IS POSITIVELY EXPANDED TO FILL THE HOLE.

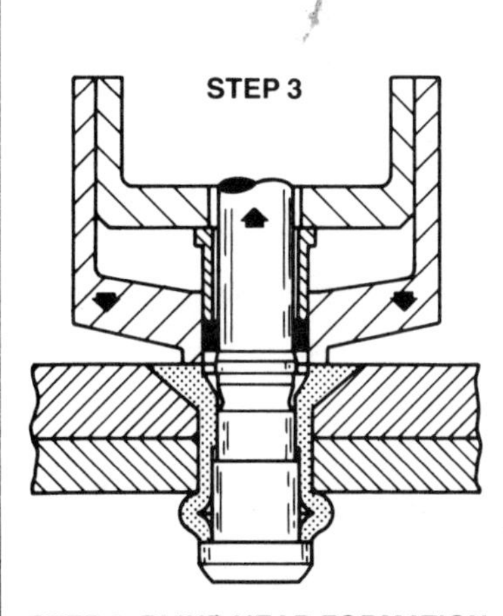

STEP 3. BLIND HEAD FORMATION• AS THE PULL CONTINUES, THE SLEEVE IS SQUEEZED BETWEEN THE HEAD OF THE SPINDLE AND THE NOSE ASSEMBLY. THE SLEEVE IS THEN UPSET TOP TO FORM A STRONG, BULBED HEAD ON THE BLIND SIDE.

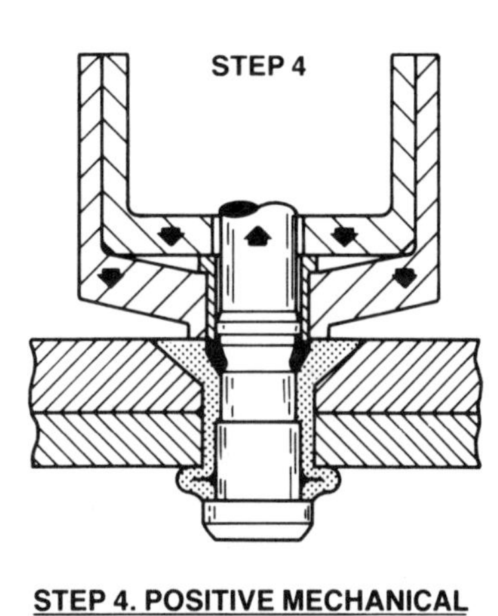

STEP 4. POSITIVE MECHANICAL LOCK• THE BLIND BULBED HEAD IS FULLY FORMED. THEN THE TOOL AUTOMATICALLY SHIFTS TO THRUST THE INNER ANVIL AGAINST THE LOCKING COLLAR WHICH IS FORCED INTO THE CONICAL RECESS. THUS THE PARTS ARE LOCKED TOGETHER PERMANENTLY.

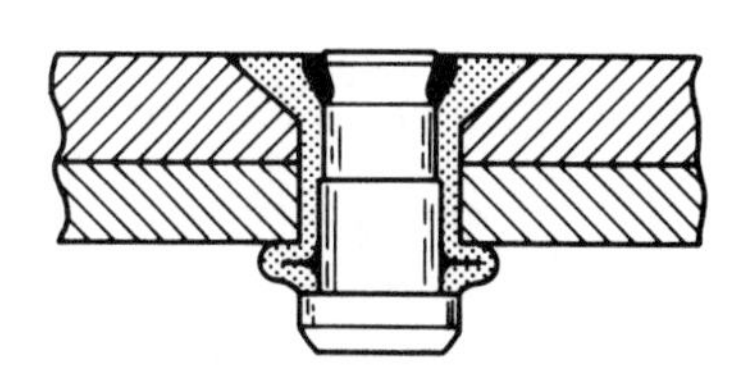

STEP 5. SPINDLE BREAK• SPINDLE IS FRACTURED IN TENSION PROVIDING A FLUSH INSTALLED BLIND RIVET WITH STRUCTURAL INTEGRITY.

Fig. 5-22 Huck fastener.

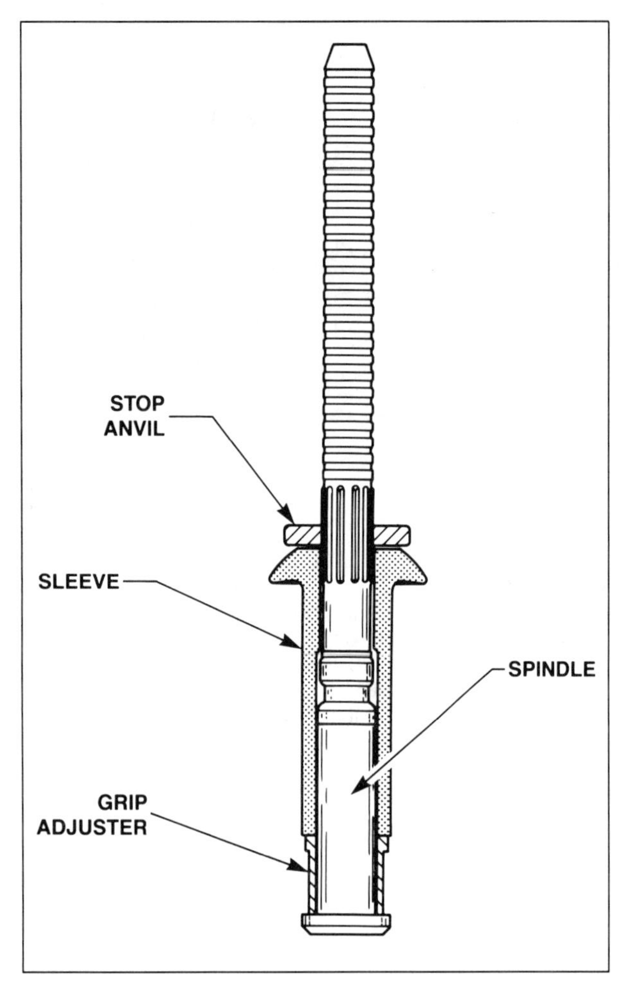

Fig. 5-23 Huck-Clinch blind rivet.

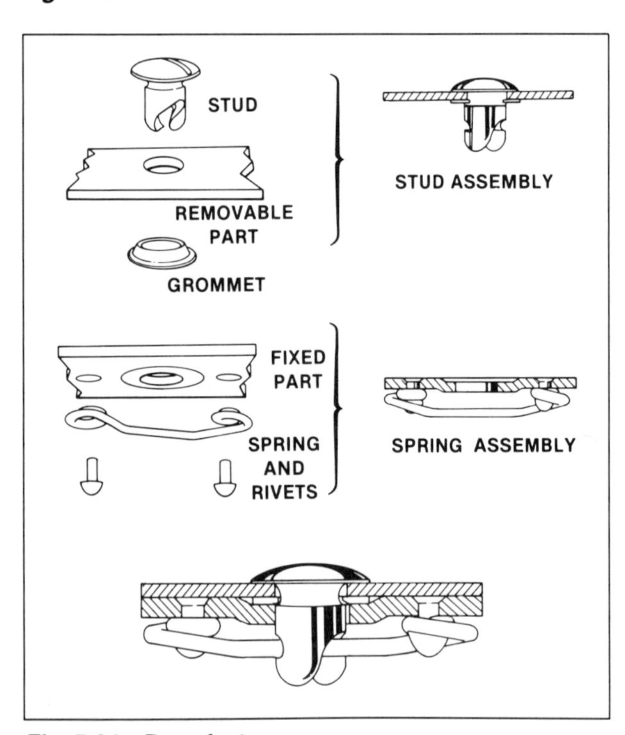

Fig. 5-24 Dzus fastener.

2. Camloc Fasteners

The ***Camloc*** fastener consists of three parts: receptacle, grommet, and stud assembly (Figure 5-25). The receptacle is made of a pressed aluminum alloy which is mounted on a stamped sheet metal base and riveted to the cowling or cowling support. The grommet is flanged to fit into a hole in the cowling. It is strengthened by ribbing that encircles the part which comes into contact with the stud.

The Camloc is designed in such a way that its crosspin can be installed by compressing it with pliers similar to those used on Clecos. To remove a Camloc receptacle, remove its two mounting rivets.

3. Airloc Fasteners

The ***Airloc*** fastener consists of two parts: spring receptacle and stud-and-crosspin assembly. The stud comes in two head styles: countersunk and round. The spring has an upper leaf that catches the crosspin. The locking of the crosspin into the upper leaf prevents the stud and pin from unlocking once it is snapped into position. The lower leaf spring forces the cowling into place. Two rivets hold the receptacle in place. The rivet heads must be countersunk so that the outer surface lies flush with the inner sheet. Figure 5-26 shows the installation of an Airloc.

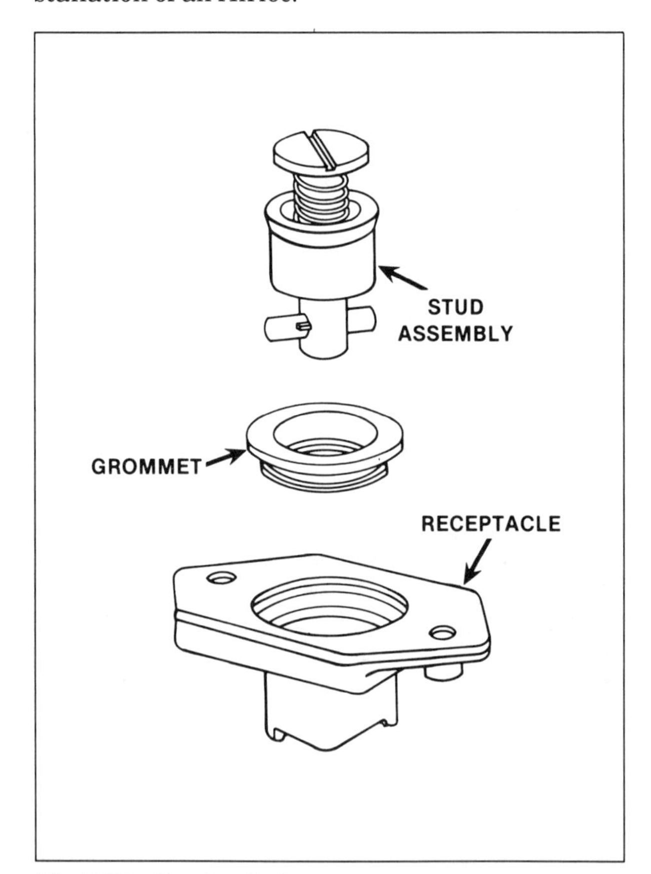

Fig. 5-25 Camloc fastener.

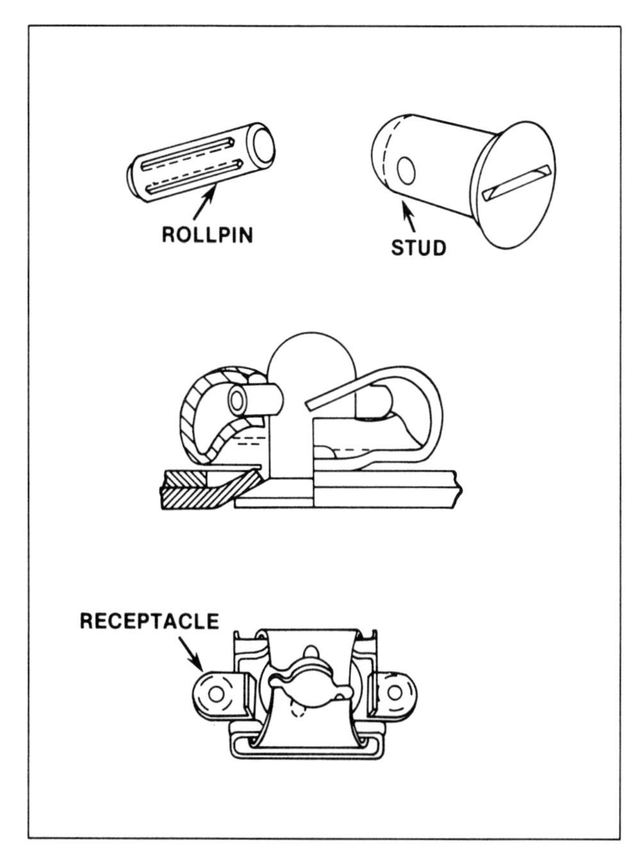

Fig. 5-26 Airloc fastener.

QUESTIONS:

1. *Name three conventional, non-blind, special fasteners.*
2. *Which is the strongest special fastener?*
3. *Name the two different types of lockbolts.*
4. *Why can a Cherrymax blind rivet be used to replace a solid-shank rivet?*
5. *Name three blind fasteners that use stem locking devices.*
6. *What three diameters are standard for most blind rivets?*
7. *What rule governs the installation of a Cherrylock friction rivet when it is used to replace a solid-shank rivet?*
8. *Which blind fastener has a locking device which is part of the manufactured rivet head?*
9. *How are blind rivets with locking stems removed?*
10. *What is the advantage of the tool used to install an Olympic-Lok rivet?*
11. *What alloys are used to make Cherrybuck fasteners?*
12. *What advantage does a Cherrybuck have over a Hi-Lok fastener?*
13. *Where are blind fasteners used most successfully?*
14. *What two head styles seem to be most common on blind fasteners?*
15. *Name three cowling and access cover fasteners.*

Chapter VI
Forming And Bend Allowance

Four steps are to be followed when forming or bending aircraft parts: Computing bend allowance and ***setback,*** measuring the layout, and making the finished part.

A. Bend Allowance

Bend allowance is defined as the amount of metal to be added to the total layout.

Figure 6-1 is an edge view of a 90° bend showing the location of bend allowance. There are three variables to consider when computing bend allowance: Radius of the bend, thickness of the metal, and number of degrees of the bend. Figure 6-2 shows a labeled, finished part.

The radius of the bend is always located at the inside heel of the bent metal. Bend radius is always expressed in thousandths of an inch. One of the

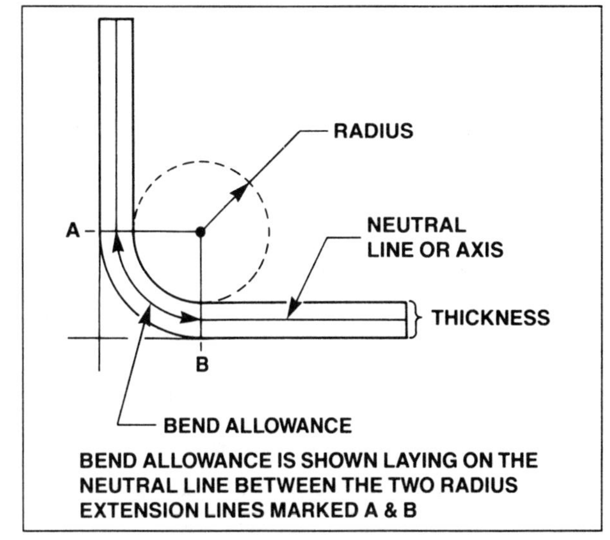

Fig. 6-1 Bend allowance.

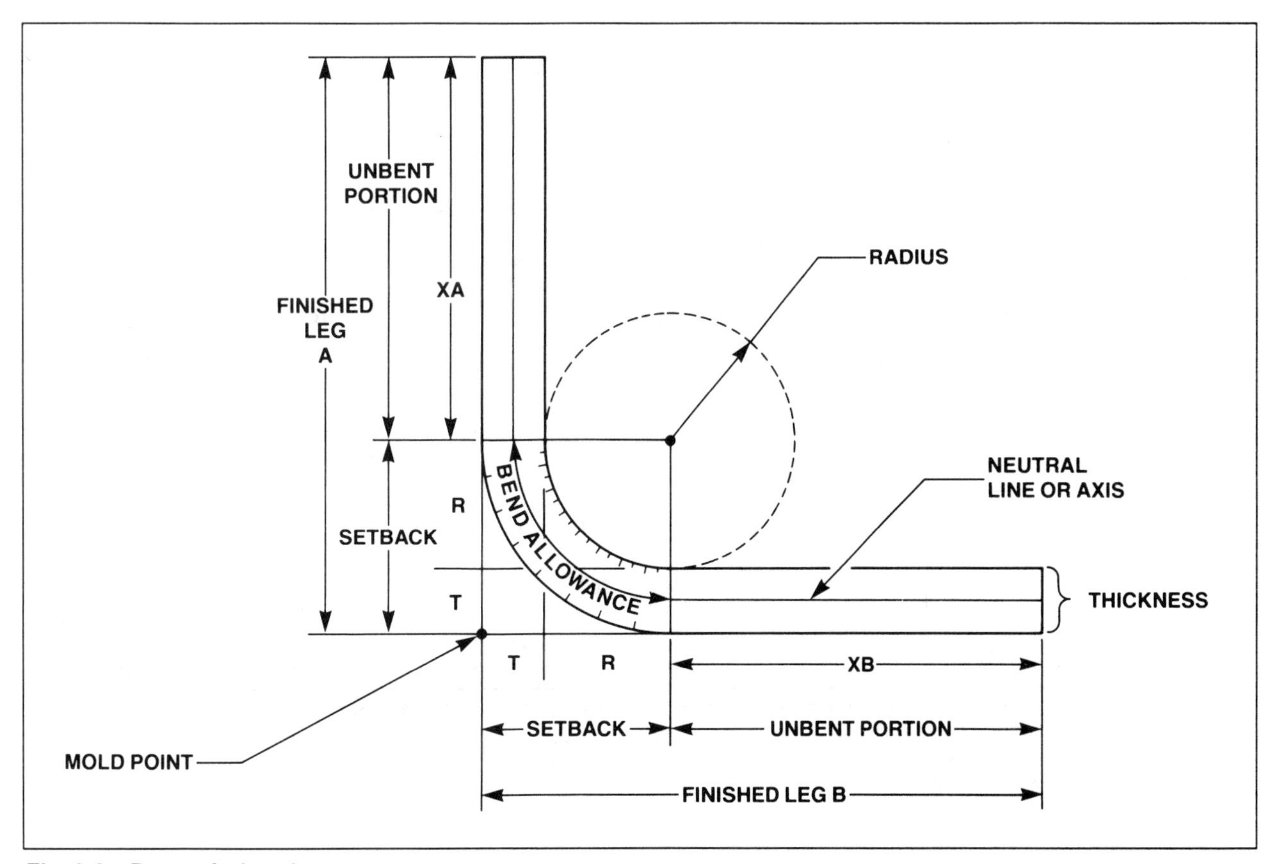

Fig. 6-2 Parts of a bend.

most important considerations when bending metal is the minimum bend radius. When bends are made smaller than the required minimum radius, metal will crack at the outside heel of the bend. A metal part bent about a radius bar will compress at the inside heel and stretch at the outside heel of the bend. This action causes the metal to become cold worked.

Because cold working can either weaken or strengthen the metal, depending upon its previous degree of hardness, determining minimum bend radius is critical.

The minimum radius of the bend is determined by the hardness and the thickness of the metal. Figure 6-3 is a minimum bend radius chart. When very hard metals (such as 7075T6), are bent using the lower side of the minimum range specified (3t), relief holes must be drilled at various intervals along the sight line.

Some aircraft aluminum alloys are extremely hard, and when they require bending to a radius smaller than minimum, they should first be annealed, a form of heat treatment that softens metal. Annealed aluminum can be bent to a minimum radius equal to its own thickness. Then after the bending is completed, the metal is re-heat treated to restore it to its previous temper condition. If heat treating equipment is not available, use a larger bend radius.

Thickness of the metal being bent is determined by where the part will ultimately be located. The thickness of the metal is always referred to in thousandths of an inch. Bend allowance and minimum bend radius are affected by the thickness of the metal.

The number of degrees of the bend is determined by the shape of the part. Different procedures apply to calculation of bends of 90° and those less and more than 90°.

Because hard metals will ***spring back*** after bending, they should be bent five to ten degrees beyond the required bend angle.

1. Bend Allowance Formula

There are four methods for finding bend allowance. Three formulas can be used. The one-pi and two-pi formulas consider the location of bend allowance to be directly on the neutral, or center, line of the thickness of the metal. The empirical formula allows for a slight inward shift of bend allowance and is therefore considered to be more accurate. A bend allowance table presents data derived from the empirical formula.

a. One-Pi Formula

The one-pi formula determines bend allowance by computing the circumference of a circle which includes the diameter of the bend plus the thickness of the metal. The formula is:

$$BA = (\pi \times D) \div (360 \div N°)$$

BA is bend allowance; Pi (π) is 3.1416; D is bend radius times two plus the thickness of metal;

RECOMMENDED RADII FOR 90° BENDS IN ALUMINUM ALLOYS						
ALLOY AND TEMPER	APPROXIMATE SHEET THICKNESS (t) (INCH)					
	0.016	0.032	0.064	0.128	0.182	0.258
2024-0[1]	0	0—1t	0—1t	0—1t	0—1t	0—1t
2024-T3[1,2]	1-1/2t—3t	2t—4t	3t—5t	4t—6t	4t—6t	5t—7t
2024-T6[1]	2t—4t	3t—5t	3t—5t	4t—6t	5t—7t	6t—10t
5052-0	0	0	0—1t	0—1t	0—1t	0—1t
5052-H32	0	0	1/2t—1t	1/2t—1-1/2t	1/2t—1-1/2t	1/2t—1-1/2t
5052-H34	0	0	1/2t—1-1/2t	1-1/2—2-1/2t	1-1/2—2-1/2t	2t—3t
5052-H36	0—1t	1/2t—1-1/2t	1t—2t	1-1/2t—3t	2t—4t	2t—4t
5052-H38	1/2t—1-1/2t	1t—2t	1-1/2t—3t	2t—4t	3t—5t	4t—6t
6061-0	0	0—1t	0—1t	0—1t	0—1t	0—1t
6061-T4	0—1t	0—1t	1/2t—1-1/2t	1t—2t	1-1/2t—3t	2-1/2t—4t
6061-T6	0—1t	1/2t—1-1/2t	1t—2t	1-1/2t—3t	2t—4t	3t—4t
7075-0	0	0—1t	0—1t	1/2t—1-1/2t	1t—2t	1-1/2t—3t
7075-T6[1]	2t—4t	3t—5t	4t—6t	5t—7t	5t—7t	6t—10t

[1]ALCLAD SHEET MAY BE BENT OVER SLIGHTLY SMALLER RADII THAN THE CORRESPONDING TEMPERS OF UNCOATED ALLOY.

[2]IMMEDIATELY AFTER QUENCHING, THIS ALLOY MAY BE FORMED OVER APPRECIABLY SMALLER RADII.

Fig. 6-3 Minimum bend radius chart.

N is the number of degrees of bend. For example, if the bend radius is .125 and the thickness of the metal is .040 then D is .290. The bend is 90°.

$$BA = \frac{3.1416 \times .290}{4} = \frac{.9110}{4} = .2277 \text{ or } .228$$

Bend Allowance is .228.

b. Two-Pi Formula

In this formula, pi is multiplied by two, and only half of the thickness is used for computation. This formula expresses the fact that bend allowance actually lies along the neutral line of the metal thickness. The two-pi formula is:

$$BA = [(2\pi)(R + \tfrac{1}{2}T)] \div (360 \div N^\circ)$$

Using the data given for the one-pi formula:

$$\begin{aligned} BA &= 2(3.1416)(R + \tfrac{1}{2}T) \div (360 \div N^\circ) \\ &= 2 \times 3.1416\,(.125 + .5 \times .040) \div (360 \div 90) \\ &= 6.2832\,(.125 + .020) \div 4 \\ &= 6.2832\,(.145) \div 4 \\ &= .227766 \text{ or } .228 \end{aligned}$$

Bend allowance is .228.

c. Empirical Bend Allowance Formula

The empirical formula is the most accurate of all the bend allowance formulas. It is based on experiments which prove that the neutral line running through the thickness of metal shifts slightly towards the inside radius when the metal is bent, thus shortening the bend allowance several thousandths of an inch. The empirical formula takes into account the shifting of the neutral line, slightly inward, when the metal is bent.

The empirical formula:

$$BA = (0.01743 \times R + 0.0078 \times T)N^\circ$$

Unlike the one- and two-pi formulas, the empirical formula assigns constants: .01743 to be multiplied by radius, and .0078 to be multiplied by thickness. These constant values are slightly smaller than would be derived by the two-pi formula. That is, if pi is multiplied by 2 and divided by 360°, the result is .017453, rather than .01743, and if pi is divided by 360°, the result is .0087, rather than .0078. An example, using the same data, is:

$$\begin{aligned} BA &= (.01743 \times .125 + .0078 \times .040)\,90^\circ \\ &= (.0021787 + .000312)\,90^\circ \\ &= (.0024907)\,90^\circ = .224 \end{aligned}$$

Bend allowance is .224.

d. Bend Allowance Chart

The bend allowance table shown in Figure 6-4 was developed using the data from the empirical formula. It relates radius to thickness. The three-place decimal is the bend allowance for 90°. The six-place decimal is the bend allowance for one degree of bend. If the six-place decimal number is multiplied by any bend angle, it will produce the bend allowance for that angle.

For example, to find the bend allowance for a 90° bend when the radius is .125 and the thickness is .040, read across the top line to .125. Then read down the thickness column until you get to .040. The three-place decimal, .224, is the bend allowance for 90° and the six-place decimal, .002493, is for one degree of bend. Bend allowance for a 45° angle is .002493 × 45, or .112.

e. Sight line

After all the measurements are laid out on the sheet metal part, sight line must be located in the bend allowance area. This is important for proper positioning of the bending brake in the bend allowance area prior to making the bend.

Location of the sight line affects the bend allowance around the bend radius bar because when metal is bent, the outside heel stretches and the inside heel compresses, causing the metal to stretch at the point of the bend. The sight line placement for bends 90° and greater is always one bend radius out from the ***bend line*** under the nose of the brake. Figure 6-5 shows the location of the sight line for bends 90° and over.

When bends are smaller than 90°, it is obvious that setting the sight line one bend radius out from the bend line will not be possible because the bend allowance area is too small. In this case, set the nose of the brake in the middle of the bend allowance area.

B. Setback

Setback is the calculation of the amount of metal to be subtracted from a leg in order to find the length of its unbent portion, also referred to as the X distance (Figure 6-6). Simply stated, setback is determined by subtracting the radius and the thickness from the length of the finished leg.

The formula for calculating setback is:

$$SB = K \times (R + T)$$

Values of K are given in Figure 6-7A. These are based on trigonometric functions. Up to 90°, the K value is from .00873 to 1. For angles above 90°, the K value is from 1 to infinity. Note that for bends of 90°, the K factor is one; that is, setback is determined by radius (R) and thickness (T).

For bends over 90°, the factor increases to infinity, but, in practice, is not significant because radius and thickness can be measured and subtracted from the length of the finished leg. However, the inclusion of the K factor in the formula allows the accurate computation of setback when radius and thickness cannot be measured, when the bend is less than 90°.

THICKNESS \ RADIUS	1/32 .031	1/16 .063	3/32 .094	1/8 .125	5/32 .156	3/16 .188	7/32 .219	1/4 .250	9/32 .281	5/16 .313	11/32 .344	3/8 .375	7/16 .438	1/2 .500
.020	.062 .000693	.113 .001251	.161 .001792	.210 .002333	.259 .002874	.309 .003433	.358 .003974	.406 .004515	.455 .005056	.505 .005614	.554 .006155	.603 .006695	.702 .007795	.799 .008877
.025	.066 .000736	.116 .001294	.165 .001835	.214 .002376	.263 .002917	.313 .003476	.362 .004017	.410 .004558	.459 .005098	.509 .005657	.558 .006198	.607 .006739	.705 .007838	.803 .008920
.028	.068 .000759	.119 .001318	.167 .001859	.216 .002400	.265 .002941	.315 .003499	.364 .004040	.412 .004581	.461 .005122	.511 .005680	.560 .006221	.609 .006762	.708 .007853	.804 .007862
.032	.071 .000787	.121 .001345	.170 .001886	.218 .002427	.267 .002968	.317 .003526	.366 .004067	.415 .004608	.463 .005149	.514 .005708	.562 .006249	.611 .006789	.710 .007889	.807 .008971
.038	.075 .000837	.126 .001396	.174 .001937	.223 .002478	.272 .003019	.322 .003577	.371 .004118	.419 .004659	.468 .005200	.518 .005758	.567 .006299	.616 .006840	.715 .007940	.812 .009021
.040	.077 .00853	.127 .001411	.176 .001952	.224 .002493	.273 .003034	.323 .003593	.372 .004134	.421 .004675	.469 .005215	.520 .005774	.568 .006315	.617 .006856	.716 .007955	.813 .009037
.051		.134 .001413	.183 .002034	.232 .002575	.280 .003116	.331 .003675	.379 .004215	.428 .004756	.477 .005297	.527 .005855	.576 .006397	.624 .006934	.723 .008037	.821 .009119
.064		.144 .001595	.192 .002136	.241 .002676	.290 .003218	.340 .003776	.389 .004317	.437 .004858	.486 .005399	.536 .005957	.585 .006498	.634 .007039	.732 .008138	.830 .009220
.072			.198 .002202	.247 .002743	.296 .003284	.346 .003842	.385 .004283	.443 .004924	.492 .005465	.542 .006023	.591 .006564	.639 .007105	.738 .008205	.836 .009287
.078			.202 .002247	.251 .002787	.300 .003327	.350 .003885	.399 .004426	.447 .004963	.496 .005512	.546 .006070	.595 .006611	.644 .007152	.742 .008243	.840 .009333
.081			.204 .002270	.253 .002811	.302 .003351	.352 .003909	.401 .004449	.449 .004969	.498 .005535	.548 .006094	.598 .006635	.646 .007176	.744 .008266	.842 .009357
.091			.212 .002350	.260 .002891	.309 .003432	.359 .003990	.408 .004531	.456 .005072	.505 .005613	.555 .006172	.604 .006713	.653 .007254	.752 .008353	.849 .009435
.094			.214 .002374	.262 .002914	.311 .003455	.361 .004014	.410 .004555	.459 .005096	.507 .005637	.588 .006195	.606 .006736	.655 .007277	.754 .008376	.851 .009458
.102				.268 .002977	.317 .003518	.367 .004076	.416 .004617	.464 .005158	.513 .005699	.563 .006257	.612 .006798	.661 .007339	.760 .008439	.857 .009521
.109				.273 .003031	.321 .003572	.372 .004131	.420 .004672	.469 .005213	.518 .005754	.568 .006312	.617 .006853	.665 .008394	.764 .008493	.862 .009575
.125				.284 .003156	.333 .003697	.383 .004256	.432 .004797	.480 .005338	.529 .005678	.579 .006437	.628 .006978	.677 .007519	.776 .008618	.873 .009700
.156					.355 .003939	.405 .004497	.453 .005038	.502 .005579	.551 .006120	.601 .006679	.650 .007220	.698 .007761	.797 .008860	.895 .009942
.188						.417 .004747	.476 .005288	.525 .005829	.573 .006370	.624 .006928	.672 .007469	.721 .008010	.820 .009109	.917 .010191
.250								.568 .006313	.617 .06853	.667 .007412	.716 .007953	.764 .008494	.863 .009593	.961 .010675

Fig. 6-4 Bend allowance chart.

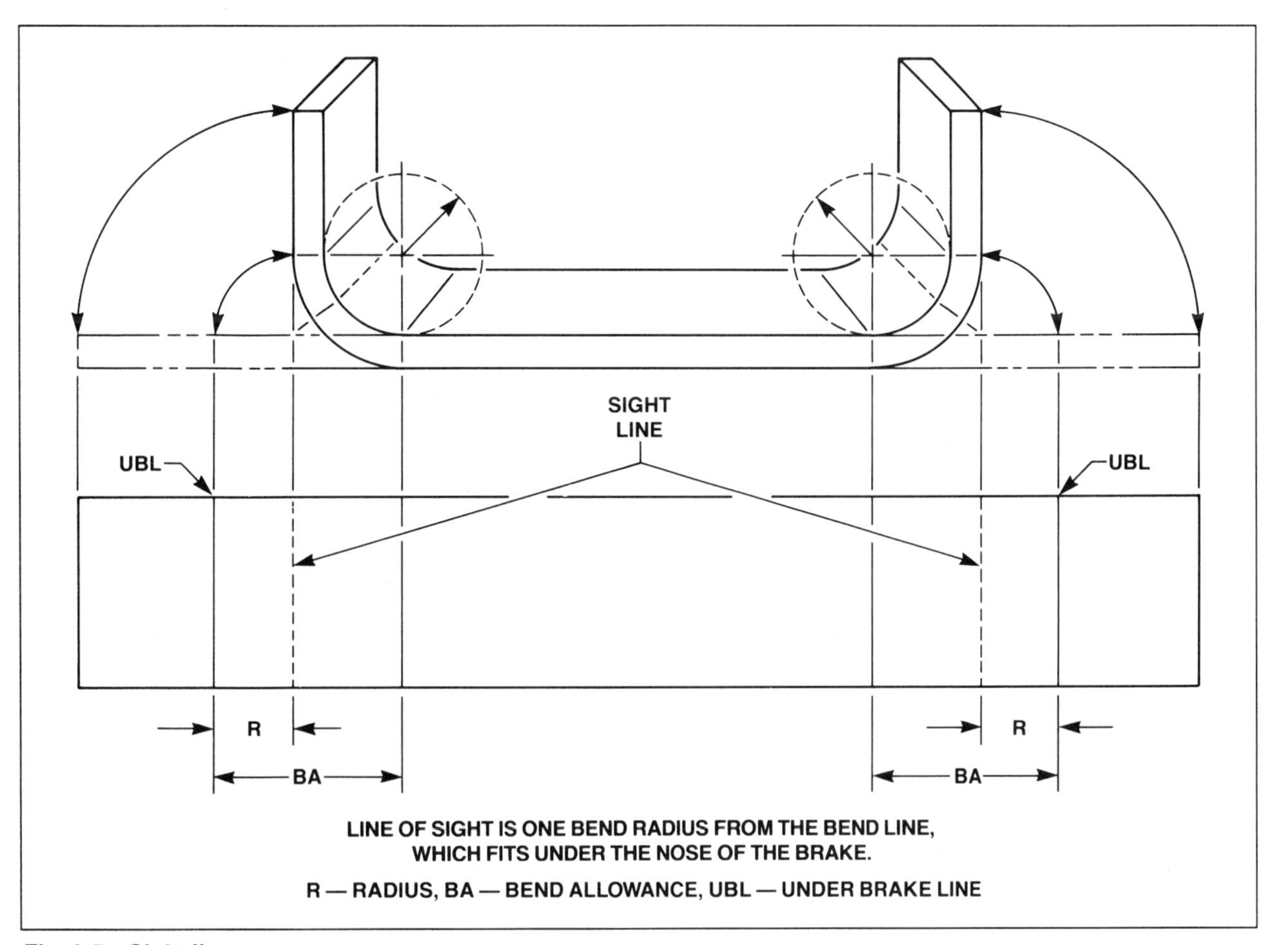

Fig. 6-5 Sight line.

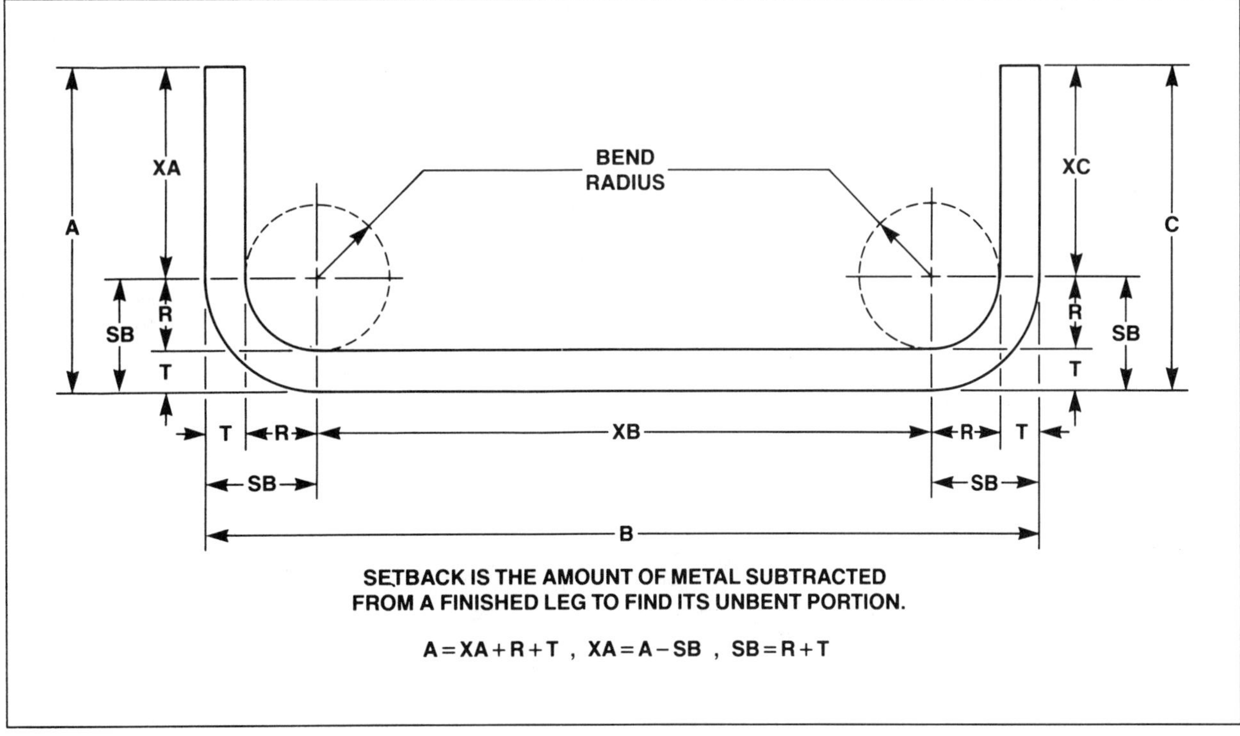

Fig. 6-6 Setback for 90° bends.

ANG [DEG]	K-VALUE	ANG [DEG]	K-VALUE	ANG [DEG]	K-VALUE
1	0.00873	61	0.58904	121	1.7675
2	0.01745	62	0.60086	122	1.8040
3	0.02618	63	0.61280	123	1.8418
4	0.03492	64	0.62487	124	1.8807
5	0.04366	65	0.63707	125	1.9210
6	0.05241	66	0.64941	126	1.9626
7	0.06116	67	0.66188	127	2.0057
8	0.06993	68	0.67451	128	2.0503
9	0.07870	69	0.68728	129	2.0965
10	0.08749	70	0.70021	130	2.1445
11	0.09629	71	0.71329	131	2.1943
12	0.10510	72	0.72654	132	2.2460
13	0.11393	73	0.73996	133	2.2998
14	0.12278	74	0.75355	134	2.3558
15	0.13165	75	0.76733	135	2.4142
16	0.14054	76	0.78128	136	2.4751
17	0.14945	77	0.79543	137	2.5386
18	0.15838	78	0.80978	138	2.6051
19	0.16734	79	0.82434	139	2.6746
20	0.17633	80	0.83910	140	2.7475
21	0.18534	81	0.85408	141	2.8239
22	0.19438	82	0.86929	142	2.9042
23	0.20345	83	0.88472	143	2.9887
24	0.21256	84	0.90040	144	3.0777
25	0.22169	85	0.91633	145	3.1716
26	0.23087	86	0.93251	146	3.2708
27	0.24008	87	0.94890	147	3.3759
28	0.24933	88	0.96569	148	3.4874
29	0.25862	89	0.98270	149	3.6059
30	0.26795	90	1.0000	150	3.7320
31	0.27732	91	1.0176	151	3.8667
32	0.28674	92	1.0355	152	4.0108
33	0.29621	93	1.0538	153	4.1653
34	0.30573	94	1.0724	154	4.3315
35	0.31530	95	1.0913	155	4.5107
36	0.32492	96	1.1106	156	4.7046
37	0.33459	97	1.1303	157	4.9151
38	0.34433	98	1.1504	158	5.1455
39	0.35412	99	1.1708	159	5.3995
40	0.36397	100	1.1917	160	5.6713
41	0.37388	101	1.2131	161	5.9758
42	0.38386	102	1.2349	162	6.3137
43	0.39391	103	1.2572	163	6.6911
44	0.40403	104	1.2799	164	7.1154
45	0.41421	105	1.3032	165	7.5957
46	0.42447	106	1.3270	166	8.1443
47	0.43481	107	1.3514	167	8.7769
48	0.44523	108	1.3764	168	9.5144
49	0.45573	109	1.4019	169	10.385
50	0.46631	110	1.4281	170	11.430
51	0.47697	111	1.4550	171	12.706
52	0.48773	112	1.4826	172	14.301
53	0.49858	113	1.5108	173	16.350
54	0.50952	114	1.5399	174	19.081
55	0.52057	115	1.5697	175	22.904
56	0.53171	116	1.6003	176	26.636
57	0.54295	117	1.6318	177	38.188
58	0.55431	118	1.6643	178	57.290
59	0.56577	119	1.6977	179	114.590
60	0.57735	120	1.7320	180	INFINITE

Fig. 6-7A K-Factor chart 1 to 180 degrees.

DEG.	SIN	TAN	COT	COS	DEG.
0.0	0.00000	0.00000	∞	1.0000	90.0
.1	.00175	.00175	573.0	1.0000	89.9
.2	.00349	.00349	286.5	1.0000	.8
.3	.00524	.00524	191.0	1.0000	.7
.4	.00698	.00698	143.24	1.0000	.6
.5	.00873	.00873	114.59	1.0000	.5
.6	.01047	.01047	95.49	0.9999	.4
.7	.01222	.01222	81.85	.9999	.3
.8	.01396	.01396	71.62	.9999	.2
.9	.01571	.01571	63.66	.9999	89.1
22.0	0.3746	0.4040	2.475	0.9272	68.0
.1	.3762	.4061	2.463	.9265	67.9
.2	.3778	.4081	2.450	.9259	.8
.3	.3795	.4101	2.438	.9252	.7
.4	.3811	.4122	2.426	.9245	.6
.5	.3827	.4142	2.414	.9239	.5
.6	.3843	.4163	2.402	.9232	.4
.7	.3859	.4183	2.391	.9225	.3
.8	.3875	.4204	2.379	.9219	.2
.9	.3891	.4224	2.367	.9212	67.1
44.0	0.6947	0.9657	1.0355	0.7193	46.0
.1	.6959	.9691	1.0319	0.7181	45.9
.2	.6972	.9725	1.0283	.7169	.8
.3	.6984	.9759	1.0247	.7157	.7
.4	.6997	.9793	1.0212	.7145	.6
.5	.7009	.9827	1.0176	.7133	.5
.6	.7022	.9861	1.0141	.7120	.4
.7	.7034	.9896	1.0105	.7108	.3
.8	.7046	.9930	1.0070	.7096	.2
.9	.7059	.9965	1.0035	.7083	45.1
45.0	0.7071	1.0000	1.0000	0.7071	45.0
DEG.	COS	COT	TAN	SIN	DEG.

Fig. 6-7B Tangent and cotangent excerpt table.

The K-factor values are arrived at by dividing any angle by two and finding what the tangent equivalent is. For example, a 45° angle divided by two is 22.5. On the tangent table, Figure 6-7B, find its value, .4142.

To determine the setback for a 45° angle with a radius of .125 and a thickness of .040, use the chart to determine the setback, .4142, and apply the formula:

$$SB = .4142 \times (.125 + .040) = .068$$

C. Layout

A layout is made by placing the measurements obtained by calculating bend allowance and setback onto sheet metal stock. To make a layout for a part with three different legs, such as a rib or spar, assign a letter to each leg (Figure 6-8). After the lines are laid out, the part is cut out and bent to shape.

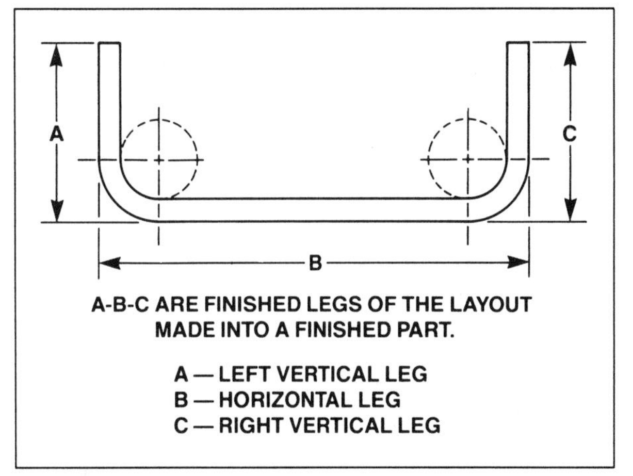

Fig. 6-8 Leg identification and names.

ITEM	FINISHED LEG	X-DIST. UNBENT LEG	BEND ALLOWANCE	COMBINED LAYOUT	FRACTION
A	1.000	.835		.835	53/64
BA			.224	.224	7/32
B	6.000	5.670		5.670	5-43/64
BA			.224	.224	7/32
C	1.000	.835		.835	53/64
T					49/64

A — LEFT VERTICAL LEG
B — HORIZONTAL LEG
C — RIGHT VERTICAL LEG

BA — BEND ALLOWANCE
T — THICKNESS

Fig. 6-9 Layout of a U-shaped channel.

1. 90° Bends

A good example of a bend of 90° is a spar splice to be inserted into a finished part (Figure 6-9). Bend a 1″ × 6″ × 1″ U-shaped channel to a radius of .125 of an inch. The metal thickness is .040 of an inch, and each leg is to be bent 90°. According to the bend allowance chart in appendix A, bend allowance is .224 of an inch. Setback for 90° is R + T: .125 + .040 = .165.

Legs A and C are one inch long. When setback, .165, is subtracted from them, the unbent portions are .835 of an inch. Leg B is six inches long. When two setbacks are subtracted, the unbent portion is 5.670 inches.

2. Bends Less Than 90°

Open-angle bends are less than 90°. The bend allowance for open-angle bends is added to the layout, as are other bend angles. However, when laying out a bend of less than 90°, setback must be calculated using the ***K factor.*** Setback is then subtracted from each finished length to find the unbent leg portion.

For bends less than 90°, setback cannot be simply R + T because there is no clear distinction between the thicknesses of the metal and the radius of the bend. An example of this method is illustrated in Figure 6-10.

3. Bends Greater Than 90°

In Figure 6-11, two identical 135°-angle bends are shown. Each of the finished parts has one-inch legs. The unbent portion of leg A in example I is obtained by subtracting the sum of R + T from the finished leg. In example II, the unbent portion of leg A is calculated by subtracting setback, which is determined by using the K factor.

The main difference between the two methods is that in example I, setback is measured from the curvature of the metal to the bend radius extension line, and in example II, setback is measured from the mold point to the bend radius extension line.

When bends are greater than 90°, bend allowance increases and setback remains the same, as it would for a 90° bend. For example, a bend over 90° on a V-shaped stringer can be calculated using this data: Bend radius is .125 inches; the thickness of the metal is .040 of an inch; the bend angle is 120°; A and B are both two inches. According to the bend allowance chart, bend allowance for one degree is .002493. Thus, BA is 120 × .002493, or .299 inches. Setback is R (.125) + T (.040), or .165 inches. Subtract setback (.165) from both A (2.000) and B (2.000) to find the unbent portion of each leg, 1.835 inches. The total layout, A + BA + B, is 3.969 inches.

D. Finished Part

The finished part is the starting point for computing bend allowance and setback. The measurements — thickness of the metal, radius of the bend, and number of degrees of the bend — are all determined from the finished part. The finished part is the actual project like a rib, spar, or stringer made in the shop. The part may not exist except on paper, blueprint, or AD compliance note.

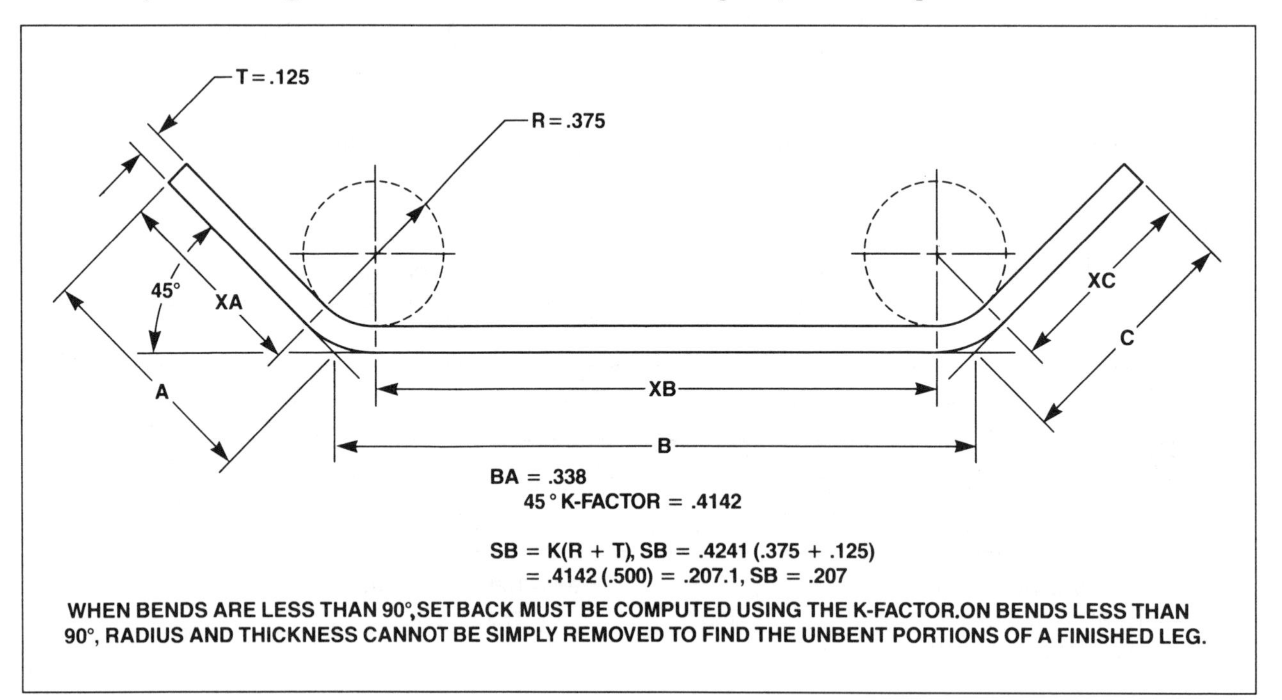

Fig. 6-10 Bends less than 90°.

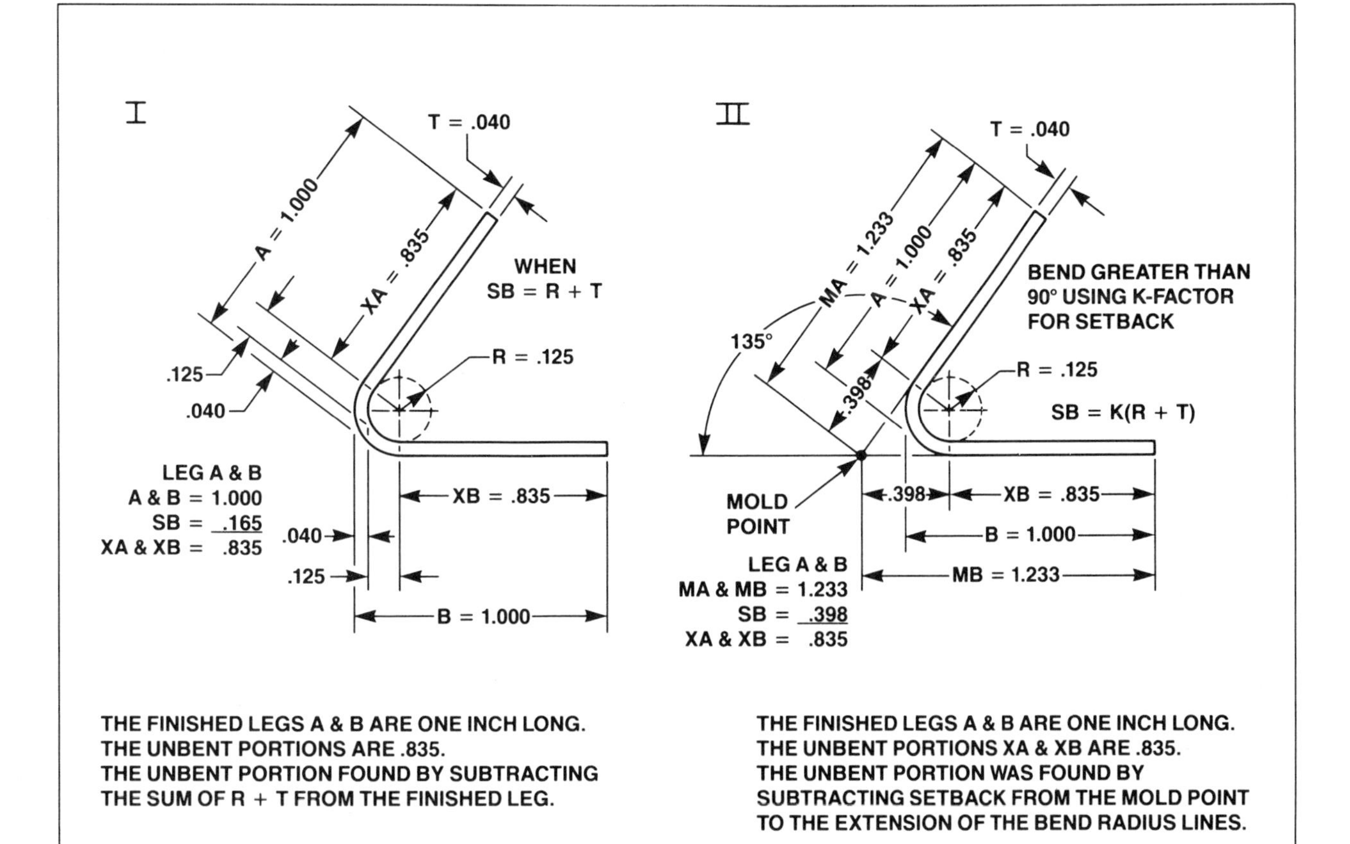

Fig. 6-11 Bends greater than 90°.

QUESTIONS:

1. *What three items must be known in order to compute bend allowance?*
2. *What happens to bend allowance when the radius is increased?*
3. *What happens to bend allowance when the thickness is decreased?*
4. *What is setback and how is it computed for a 90° bend?*
5. *On a piece of paper, do this layout: 1" × 4" × 1" U-shaped channel; bend radius is .156 inches; thickness of metal is .051 of an inch; and each bend angle is 90°.*
6. *Using the data from problem 5, compute the layout for a stringer that has an angle of 135°, if leg A = 1.000 inch and leg B = 1.750 inches.*
7. *Using the empirical formula, find bend allowance if R = .167, T = .034, and N = 90°.*
8. *Using the data from problem 7, find bend allowance by using the two-pi formula.*
9. *Define an open-angle bend.*
10. *How was the data in the K factor chart derived?*
11. *What determines minimum bend radius used when bending metal?*
12. *Where is the neutral axis on a piece of metal found?*
13. *Where does bend allowance physically occur on a bent piece of metal?*
14. *What happens to the outside heel of a bend when the metal is bent?*
15. *What happens to the inside heel of the bend when the metal is bent?*

Chapter VII
Aircraft Repairs

All major repairs must conform to the guidelines of the original certification of an aircraft, and they must meet with the approval of the FAA and the manufacturer. Before making any major structural repair or alteration, the technician should review acceptable FAA data, FAA-approved technical data, and manufacturer's recommendations.

A. Acceptable And Approved FAA Repair Data

FAA-approved technical data, the technical data which must be used in making an FAA-approved repair, includes the following: Aircraft Type Certificate (ATC) data sheets, aircraft specifications, Supplemental Type Certificates (STC), Airworthiness Directives (ADs), and manufacturers' ***FAA-approved data*** (DOA). Approved data can also be obtained from a designated engineering representative (DER) and a designated alteration station (DAS) with an FAA field approval.

When FAA-approved technical data is not available, the following technical information may be used in making major repairs or alterations: FAA Advisory Circulars (AC) 43.13-1A and 2A, manufacturers' technical information such as manuals, bulletins, kits etc., and military technical orders.

Major repairs can be made using the AC43.13-1A as approved data when the mechanic determines that it is directly applicable to the repair, and if the information is not opposed to the manufacturer's data. The ***AC 43.13-1A*** contains a series of sheet metal structural repairs which serve as general guidelines. Many of these repairs are similar to the repairs shown in the manufacturers' maintenance manuals. A good suggestion is to use the examples shown in AC 43.13-1A as a minimum standard for performing an FAA-approved repair.

B. Ordering Parts

Although parts for older aircraft may have to be made, it is more economical to buy new aircraft parts. When ordering new parts, the serial number of the aircraft must be known because later models may be slightly different in design.

In the parts book, later-model aircraft have a "usable on" code to indicate the manufacturer's changes.

An example of the code is shown in Figure 7-1. Note three different styles of wing tips: B, C, and D. Each letter represents the serial number range of the aircraft on which the wing tips are used.

C. Analyzing The Repair Area

The use of the manufacturer's maintenance and parts manual is important in determining the extent of the damaged area. List the new or rebuilt parts needed. Determine if the damage is serious enough for complete replacement of a wing or control surface. For example, it may be more economical to replace a whole wing rather than pay for replacement parts and labor.

When planning repairs, consideration must be given to the speed of the aircraft. For example, a low-speed aircraft uses surface patches or common lap joints with universal head rivets, while high-speed aircraft use flush patches and countersunk rivets.

The cost of labor and parts must be estimated and the owner so informed. Do not begin any repair work on an aircraft without the consent of the aircraft owner.

Damages that often appear on aircraft structures are ***oil canning,*** ruptures, cracking and attrition. Oil canning is caused by the loosening of skins between two ribs or stringers. The metal pops back and forth until it becomes excessively coldworked and cracks. Ruptures are caused by the exertion of force on the structural skins of light, pressurized aircraft.

Many aircraft will develop cracks near joints or seams. These cracks are usually caused by over coldworking the metal. Attrition is the failure of aircraft structures due to age and use. As an aircraft gets older, its ability to withstand the constant expansions and contractions caused by flying or landing is impaired.

In their earlier stages, many structural problems can be prevented by the addition of a stiffener in the area where the coldworking action is taking

MODEL 150 & A150

FIGURE AND INDEX NO.	PART NUMBER	DESCRIPTION 1 2 3 4 5 6 7	UNITS PER ASSY	USABLE ON CODE
5 -45	0420015-1	BRACKET-FLAP ACTUATOR MOUNT UPPER RH ONLY	1	
-46	0420015-2	BRACKET-FLAP ACTUATOR MOUNT LOWER RH ONLY	1	
-47	0426515-2	BRACKET-FLAP BELLCRANK SUPPORT UPPER LH	1	
	0426515-3	BRACKET-FLAP BELLCRANK SUPPORT UPPER R	1	
-48	0426515-4	BRACKET-FLAP BELLCRANK SUPPORT LOW	1	
	0426515-5	BRACKET-FLAP BELLCRANK SUPPORT	1	
-49	0422408-1	PLATE-REINFORCEMENT	1	
-50	0523219-2	CHANNEL-AILERON BELLC	1	
	0523219-3	CHANNEL-AILERON B	1	
-51	0523219-4	CHANNEL-AILER	1	
	0523219-5	CHANNEL-	1	
-52	0422413-1	STRI	1	
	0422413-2		1	
-53	0422413-3		2	
	0422413-4		2	
-54	NAS680A		2	
-55	04		1	
			1	
-56			2	
			2	
-57			2	
-58			1	
			1	
-59			1	
-60	C		1	
-61	0		1	
-62	04		1	
	04		1	
-63	042		1	
-64	042		1	
-65	0422		1	
	04224	RH	1	
-66	04224		1	
-67	S1209-	WING	1	
	NAS680A		6	
-68	S237-2	NG	3	
-69	NAS395-1		3	
-70	S330-2	ION OPENING	3	
-71	NAS395-14-		4	
-72	CM2692-28-	UTTON	1	
-73	0721107-1	AIRING LH	1	
	0721107-2	FAIRING RH	1	
-74	0426209-2	DOOR CATCH ASSEMBLY	1	
		ATTACHING PARTS		
	AN515-6R6	SCREW	1	
	NAS680A06	NUTPLATE	1	
		---*---		
-75	0423008-3-791	TIP-WING STANDARD	2	B
-76	0523565-201-791	TIP-WING LH CONICAL CAMBER	1	C
	0523565-200-791	TIP-WING RH CONICAL CAMBER	1	C
	0523565-29-791	TIP ASSEMBLY-WING LH CONICAL CAMBER	1	D
	0523565-30-791	TIP ASSEMBLY-WING RH CONICAL CAMBER	1	D
		ATTACHING PARTS		
	S1021Z8-8	SCREW	12	
	0423008-4	STIFFENER USE ON UPPER SIDE LH TIP & LOWER SIDE RH TIP USED WITH STANDARD WING TIP	1	B
	0423008-5	STIFFENER USE ON UPPER SIDE RH TIP & LOWER SIDE LH TIP USED WITH STANDARD WING TIP	1	B
	0523565-21	STIFFENER USE ON UPPER SIDE LH TIP & LOWER SIDE RH TIP USED WITH CONICAL CAMBER WING TIP	1	
	0523565-22	STIFFENER USE ON UPPER SIDE RH TIP & LOWER SIDE LH TIP USED WITH CONICAL CAMBER WING TIP	1	
		---*---		
	NAS696A06L	NUTPLATE	2	D
	S1860-2	NUTPLATE	3	D

A---A150 SERIAL A15000001 & ON
FA150 SERIAL FA15000001 & ON
B---A150 SERIAL A15000001 & ON
C---A150 SERIAL A15000001 THRU A15000226
FA150 SERIAL FA15000001 THRU FA15000081
D---A150 SERIAL A15000227 & ON
FA150 SERIAL FA15000082 & ON

NOTE USABLE ON CODE:
AIRCRAFT SERIAL
NUMBER IS
FA15000085 (CODE-D)

Fig. 7-1 Page from illustrated parts catalog.

place. Excessive coldworking faults are commonly associated with light, thin-gauge aircraft skins. Larger aircraft are also affected over a longer period of time.

D. Removal Of The Damaged Parts

Cleaning out the damaged area involves the removal of all bent or broken parts. Disassemble the aircraft carefully because many of the parts may be re-usable or repairable. As the parts are removed, they should be identified by part and re-installation number with a permanent felt marker. The permanent felt markings can later be washed away with alcohol.

Warning: When removing rivets, never oversize the holes.

The recommended procedure for rivet removal is to use a drill, one size smaller than the rivet being removed. Drill only the depth of the manufactured head. Use a pin punch the same size as the rivet, and snap the drilled heads off. Back up the shop head side of the rivet shank and tap out the remaining stem.

E. Installation Of New Or Rebuilt Parts And Patches

Repair parts must be as strong as the original. The metal used to make the parts must be the same alloy content and thickness as the original. The parts must be re-installed using the same size and kind of rivets as the original design.

A hole finder will be needed to locate the centers of new rivet holes when new skins are being prepared for installation (Figure 7-2). When re-assembly begins, do not rivet anything in place until all parts are fitted together and the holes line up. Hold the work in place with Clecos placed two or three inches apart. Be sure that all pilot holes are brought up to specifications before riveting begins.

A patch on the leading edge of a light aircraft wing flap serves to restore its original strength and shape. An example of a repair to a flap leading edge

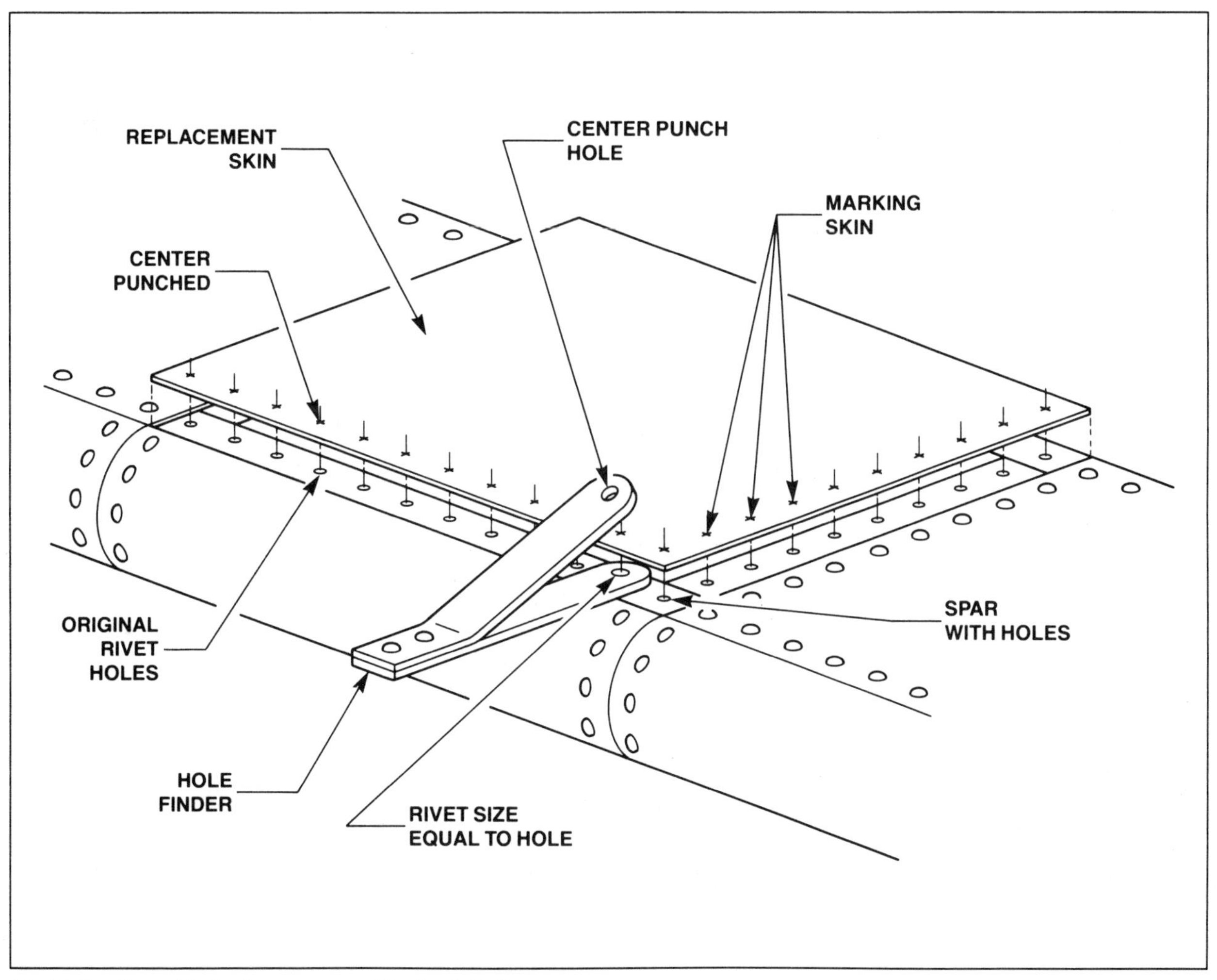

Fig. 7-2 New skin being prepared for installation.

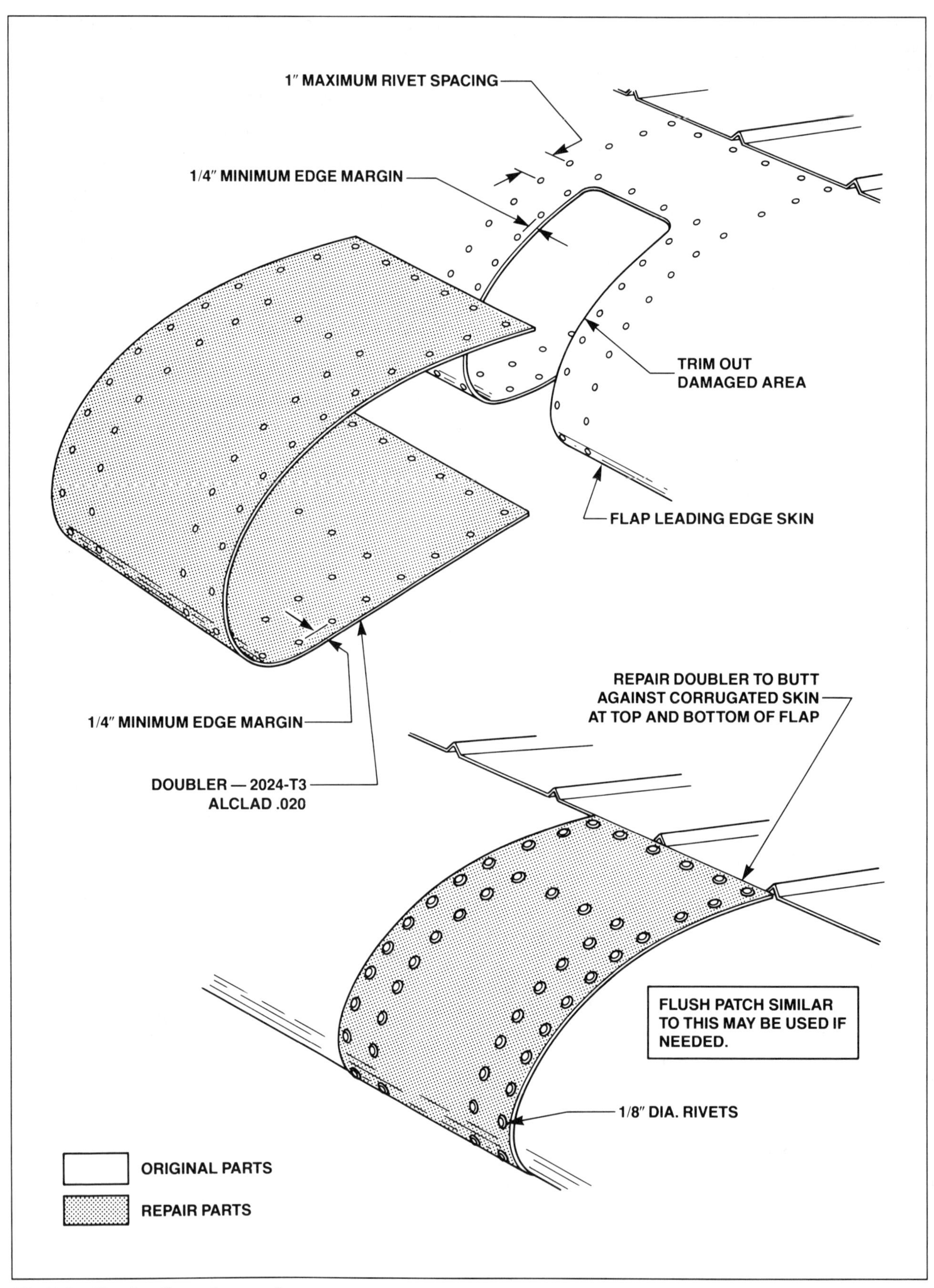

Fig. 7-3 Flap leading edge repair.

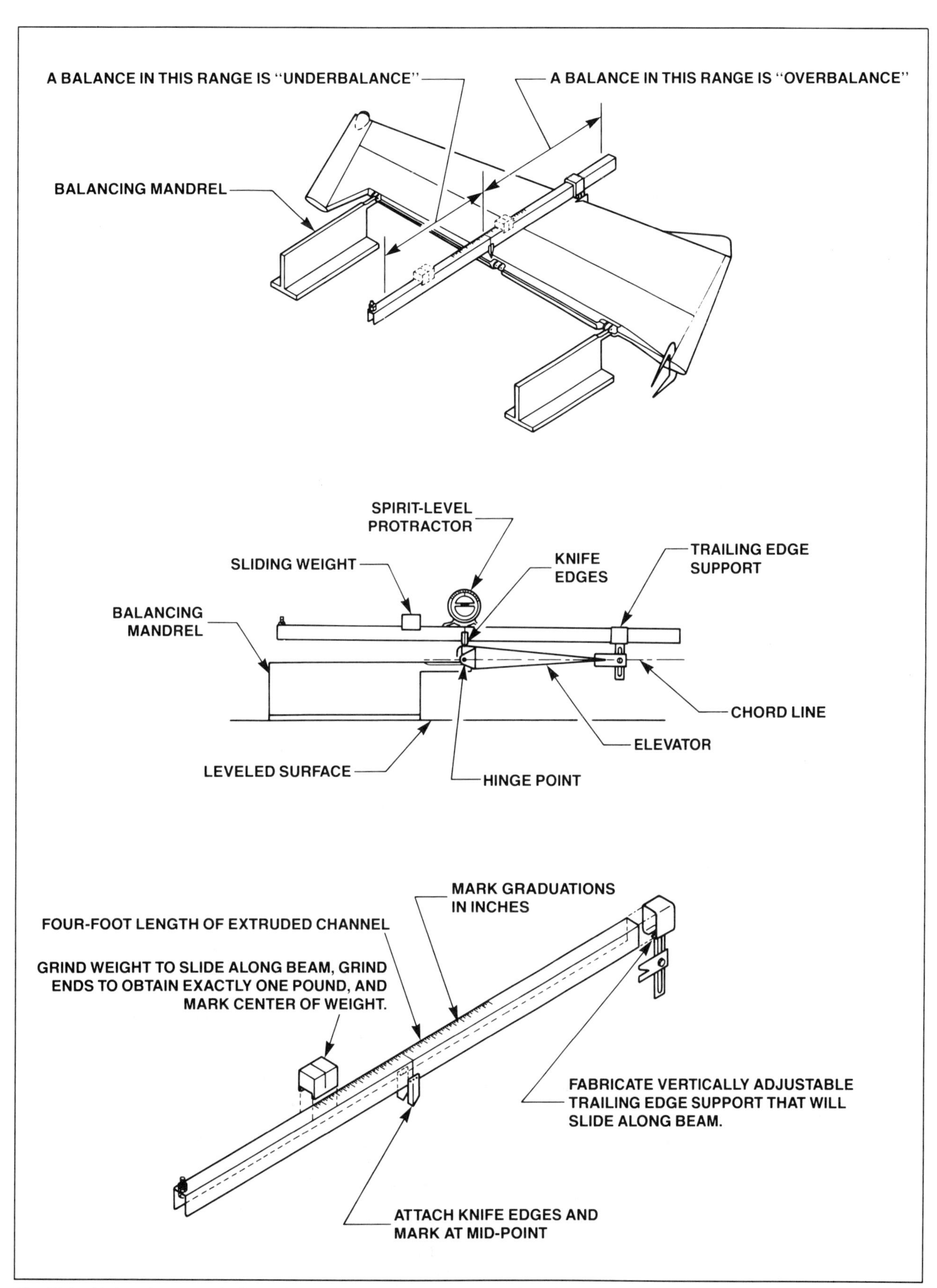

Fig. 7-4 Balancing mandrel and beam for re-balancing control surfaces.

requiring the removal of the damaged area is shown in Figure 7-3. The clean-out must be held to a minimum to make the patch as small as possible in order to save weight.

After a primary control surface (aileron, elevator, or rudder) is repaired, it must be re-balanced. If a control surface is not in balance, it will cause flutter and eventually lead to serious structural damage at its connecting points. See Figure 7-4 for examples of the tools and devices used for re-balancing control surfaces. Consult the manufacturer's maintenance manual for the specifications.

1. Surface And Flush Patches

Two types of patches are used on aircraft: surface patch and flush patch (Figure 7-5). These patches are ordinarily round, oval, or square with 1/2-inch radius corners. When patches are made on wings, fuselage, or control surfaces, a prescribed number of rivets must be used to obtain the necessary bearing and shearing strengths.

F. Rivet Formulas

The formula for finding rivets per inch is:

$$RPI = T \times 75000 \div DSS$$

T is the thickness of the metal; DSS is the driven shear strength of the rivet (see Appendix A-V for the rivets-per-inch chart, based on this formula). For example, if the DSS is 389 and the metal is .040 inch thick, the number of rivets per inch is 7.7.

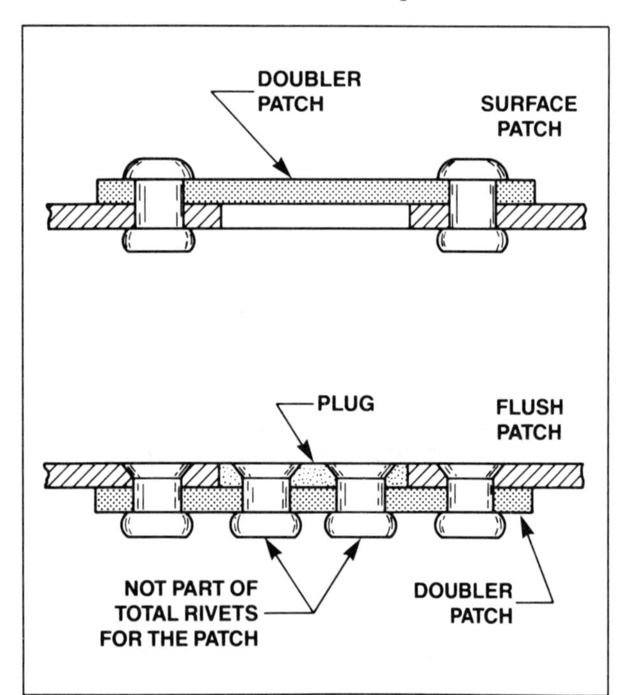

Fig. 7-5 Surface and flush patches.

To determine the total number of rivets for a patch, use the following formula:

$$NR = (L \times RPI \times \%)2$$

NR is the number of rivets; L is the length of the break; RPI is rivets per inch, as given in the chart. The % in the formula for a lap joint or common patch is 75%. For example, if a patch is covering a 3-inch diameter hole using AN470AD4 rivets, the total number of rivets is 36.

$$\begin{aligned} NR &= (3 \times 7.7 \times .75)2 \\ &= (17.325)2 \\ &= (18)2 \\ &= 36 \end{aligned}$$

The total number of rivets is 36.

A new, simplified rivet formula can be used to find the total number of rivets needed for a patch when the chart cannot be used:

$$NR = (L \times T \times 75000 \div DSS) \times \% \times 2$$

NR is the number of rivets; L is the length of the break; T is the thickness of the metal; 75,000 is the skin stress; DSS is the driven shear strength; and 75% is the percentage of rivets used for a lap joint. For example, if the thickness of the metal is .040, DSS is 389, and % is .75. The total number of rivets is 36:

$$\begin{aligned} NR &= (3 \times .040 \times 75000 \div DSS)(.75)2 \\ &= (9000 \div 389)(.75)2 \\ &= (17.352)2 \\ &= (18)2 \\ &= 36 \text{ rivets} \end{aligned}$$

Either the rivets-per-inch chart (Appendix A-V) or the simplified formula may be used to make a patch repair on an aircraft because they yield the same results. The chart can only be used for universal head rivets. The formula can be used for either universal or countersunk rivets.

1. Surface Patch Layout

Figure 7-6 shows the layout of a surface patch covering a one-inch hole. This patch uses eight each AN470 AD4 rivets spaced 45° apart. The overall diameter of the patch is 2.50 inches. The edge distance is 3 rivet diameters.

Figure 7-7 shows the layout of a surface patch covering a two-inch hole. This patch uses sixteen each AN470AD4 rivets spaced 45° apart in each row. Edge distance is four rivet diameters. A doubler is installed to increase the skin strength in the area of the patch. The total size of this patch is 5 inches.

Figure 7-8 shows the layout of a surface patch covering a three-inch hole. This patch uses 24 each AN470AD4 rivets spaced 30° apart in each row.

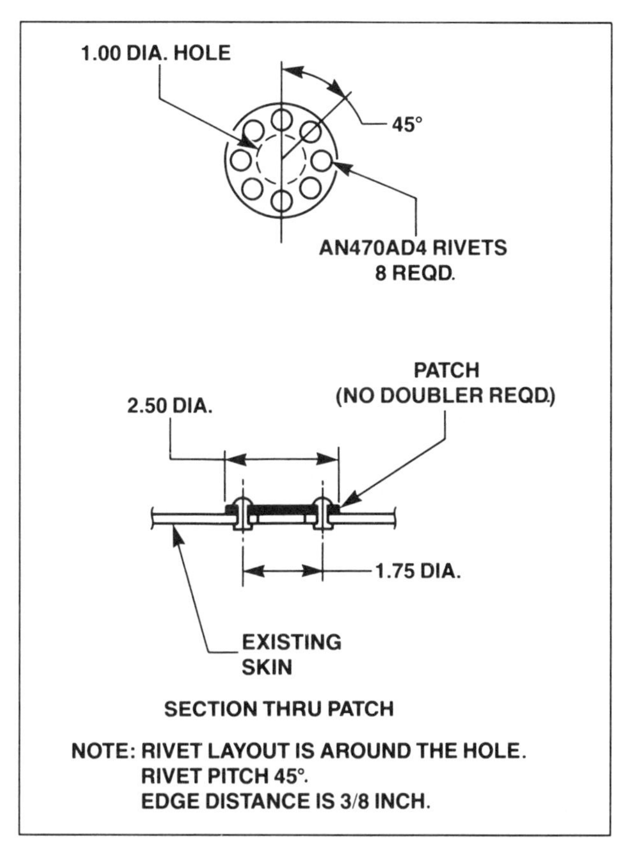

Fig. 7-6 Patch repair for one-inch diameter hole.

Edge distance is 4 rivet diameters. A doubler was installed on the inside of the hole to add strength to the skin in the area. The size of this patch is 7.5 inches.

These examples show that the number of rivets increases as the size of the repair increases. The total number of rivets used in each case can be arrived at by using the rivet-per-inch chart (Appendix A-V) or the new, simplified rivet formula. An example using the table is:

NR = (1 x 4.9 x .75)2
= (3.675)2
= (4)2
= 8 rivets

The total number of rivets for the patch is 8.

The new, simplified rivet formula can be applied to the same data:

NR = (1 x .025 x 75000 ÷ 389) (.75)2
= (1875 ÷ 389) (.75)2
= (4.82) (.75)2
= (3.6)2
= (4)2
= 8 rivets

The total number of rivets for the patch is 8.

Figure 7-9 shows an alternative layout for a rivet patch not using angular rivet spacing. The following facts are known: The patch uses 24 each AN470AD4

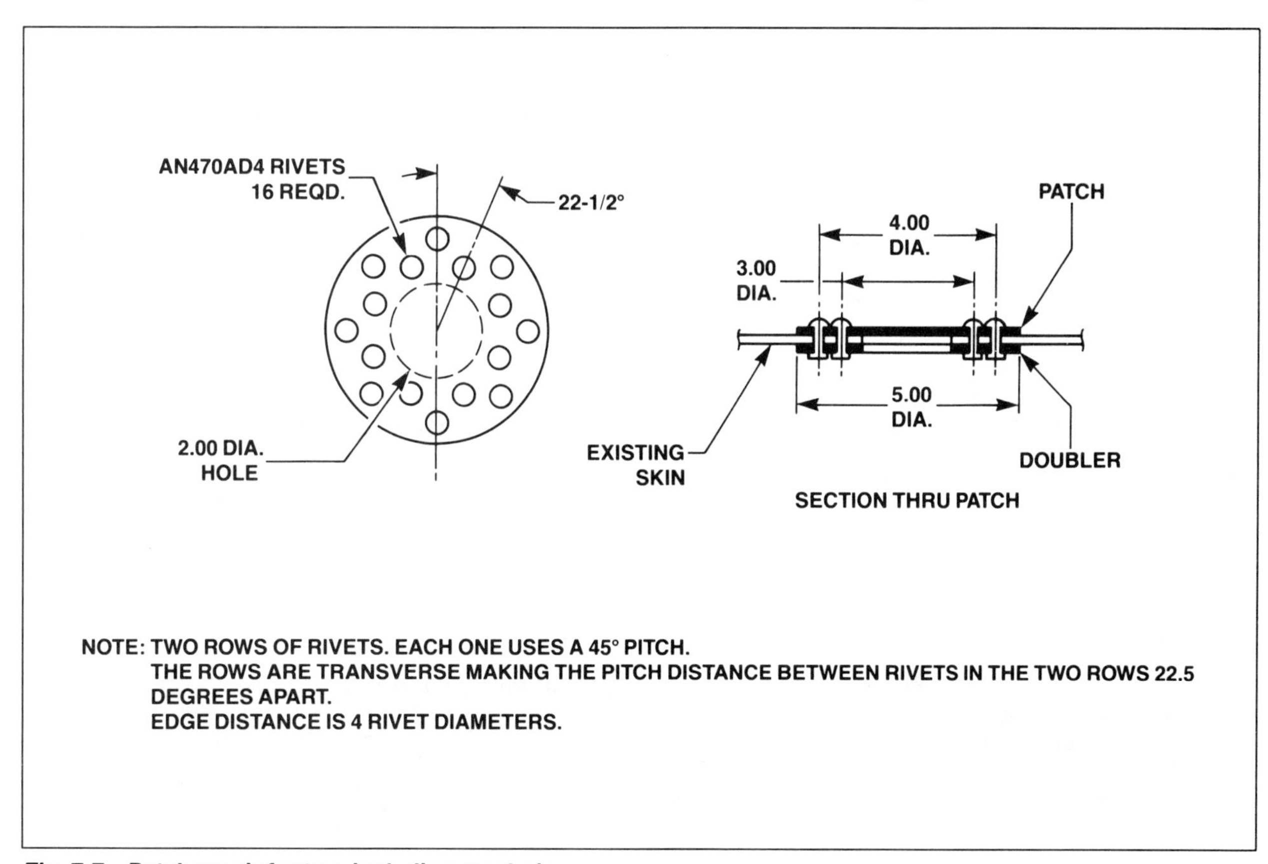

Fig. 7-7 Patch repair for two-inch diameter hole.

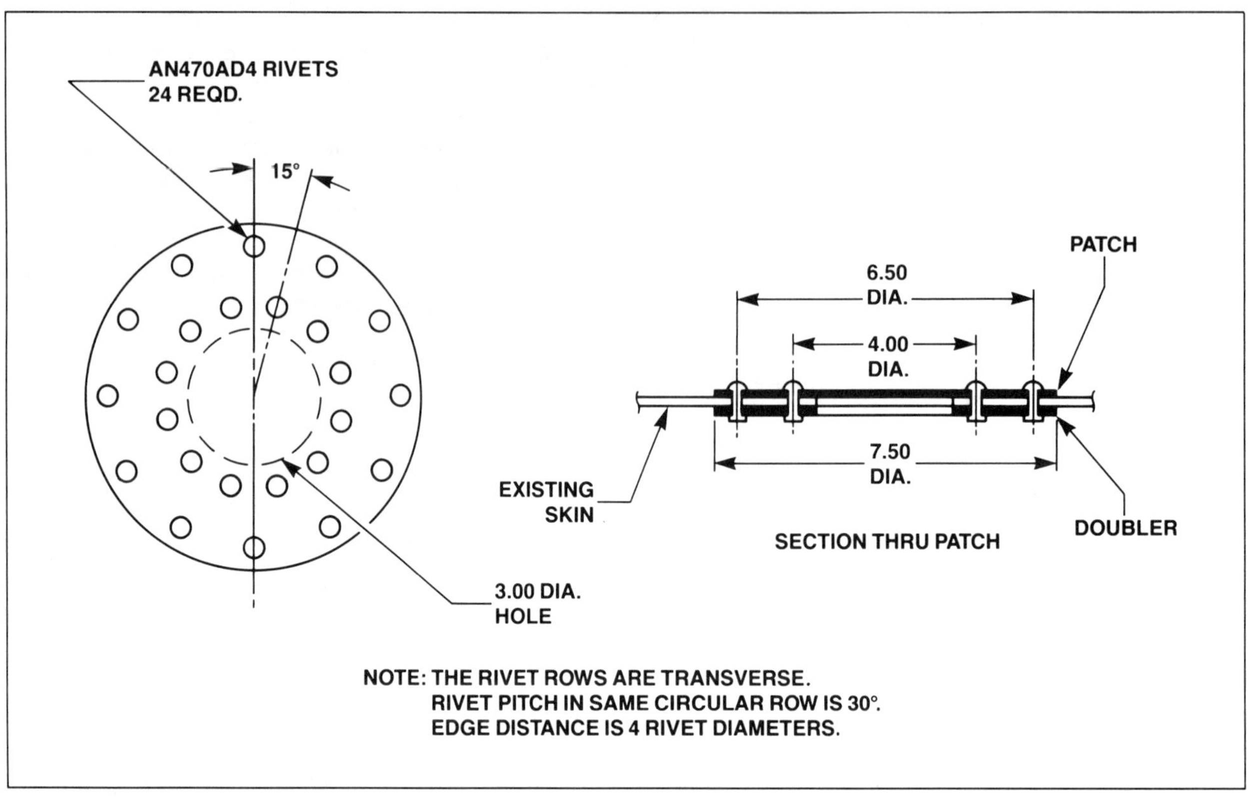

Fig. 7-8 Patch repair for three-inch diameter hole.

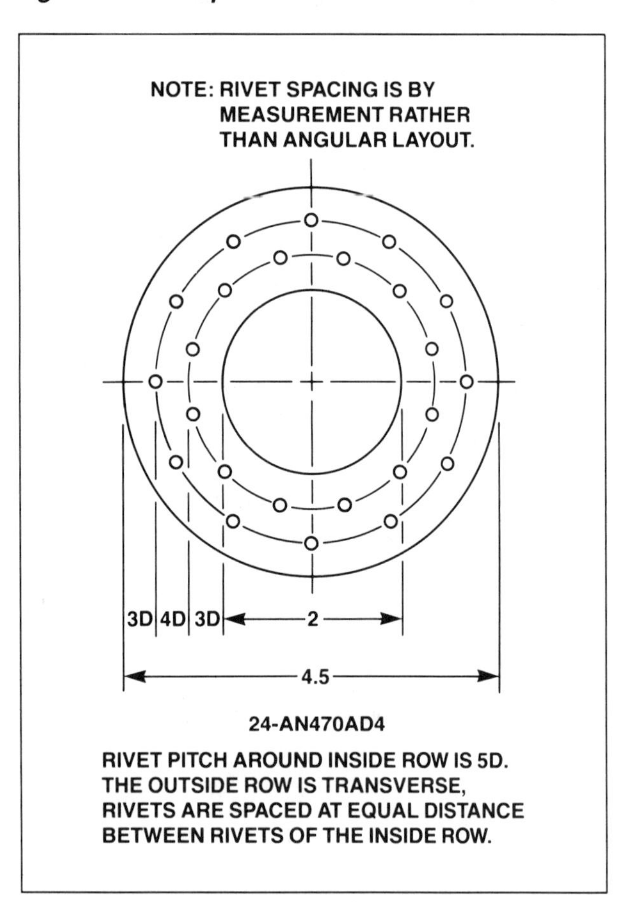

Fig. 7-9 An alternative patch method.

rivets; there are 12 rivets in each of two rows; the hole being patched is 2 inches in diameter; edge distance is 3 rivet diameters; the circumference of the first row of rivets is 7.4613".

To determine the approximate spacing of the 12 rivets in the first row, divide the circumference (7.4613) by 12. Each space will be slightly larger than 5/8 of an inch. The outside row of rivets will be located 4 rivet diameters from the inside row and will be spaced transverse to it. The patch will end with an edge distance of 3 rivet diameters from the outside row. The finished patch has a diameter of 4.5 inches.

2. Flush Patch Layout

Flush patches, made on smooth aerodynamic surfaces, use countersunk rivets (Figure 7-10). The driven shear strength of a countersunk rivet installed into a machined hole is less than that of a universal or cold-dimpled rivet. Therefore, more will be needed. For example, the driven shear strength of an AN426AD4 rivet is 331 LB. Using the simplified rivet formula:

$$\begin{aligned} NR &= (2 \times .040 \times 75000 \div 331)(.75)2 \\ &= (6000 \div 331)(.75)2 \\ &= (18.126)(.75)2 \\ &= (13.59)2 \\ &= (14)2 = 28 \text{ rivets} \end{aligned}$$

For the flush patch 28 rivets are required.

3. Access Plate Installation

Access plates are used to open up an area on a wing or fuselage in order to buck rivets on a patch or part. Access plate installation is shown in Figure 7-11. Access holes and covers are provided by the manufacturer, or they can be made in the shop. In any case, follow the instructions set forth by the manufacturer in the maintenance or part manuals when installing access plates and covers.

Access plates are always located on the underside of wings or control surfaces. They are installed near a spar, rib, or stringer in order to tie into their strength.

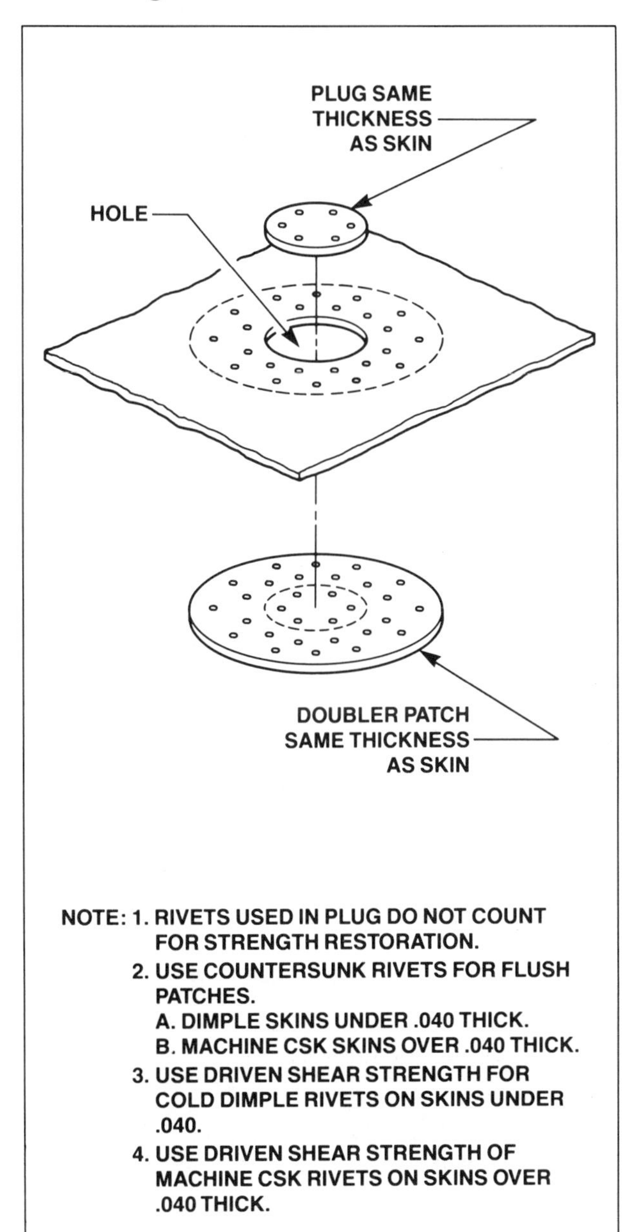

Fig. 7-10 Flush patch with countersunk rivets.

4. Corrugated Skin Repairs

Many light aircraft use ***corrugated skins*** to give the control surface rigidity and strength yet keep it light. Any repair to a corrugated control surface is considered a major repair because the control surface is balanced and the skin is considered a primary structure. Figure 7-12 shows a corrugated control surface repair.

5. Splicing A Single-Channel Spar

A repair to a single-channel spar is considered a major repair. Whenever a section is cut out of a spar in order to make a repair, it must be replaced with metal of the same thickness and strength as the original. A doubler which connects the insert to the walls of the spar can be made of thicker metal than the original.

An example of this type of repair is shown in Figure 7-13. The repair has five parts: an insert 4.5 inches high with 7/8-inch legs, a rear and a front doubler, and two 7/8″ × 7/8″ × .064″ angle stringers. Fifty-two MS20470AD4 rivets are needed to make this repair. The rivet spacing and edge distances are noted in the illustration. The driven shear strength of each rivet is 389 pounds. The total shear strength is 389 × 52, or 20,228 pounds. The rivets (12) installed into the insert are not part of the total shear strength of the joint. Their only purpose is to serve as a filler between the two doublers.

G. Inspection And Return To Service

After a major repair or alteration is completed, a final inspection must be made by an authorized FAA inspector. The inspection, which is based on manufacturer's guidelines, includes a detailed examination of the parts used and of the installation of rivets and other fasteners.

When the final inspection is completed, an ***FAA 337 form*** must be filled out in duplicate and mailed to the local FAA General Aviation District Office (GADO) within 48 hours of its being dated. The law requires that two 337 forms be made out and distributed, one to the FAA GADO and the other to the owner of the aircraft. Some mechanics make a third copy for their personal files.

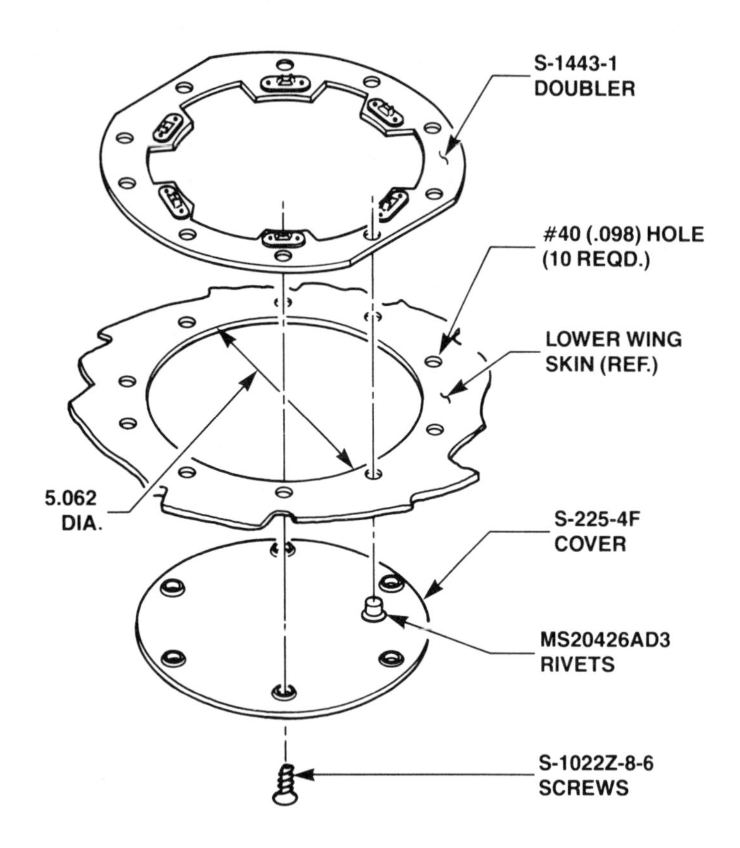

PRECAUTIONS

1. ADD THE MINIMUM NUMBER OF ACCESS HOLES NECESSARY.
2. ANY CIRCULAR OR RECTANGULAR ACCESS HOLE WHICH IS USED WITH APPROVED OPTIONAL EQUIPMENT INSTALLATIONS MAY BE ADDED IN LIEU OF THE ACCESS HOLE ILLUSTRATED.
3. USE LANDING LIGHT INSTALLATIONS INSTEAD OF ACCESS HOLES (THRU 1970 MODELS) WHERE POSSIBLE. DO NOT ADD ACCESS HOLES AT OUTBOARD END OF WING; REMOVE WING TIP INSTEAD.
4. DO NOT ADD AN ACCESS HOLE IN THE SAME BAY WHERE ONE IS ALREADY LOCATED.
5. LOCATE NEW ACCESS HOLES NEAR THE CENTER OF A BAY (SPANWISE).
6. LOCATE NEW ACCESS HOLES FORWARD OF THE FRONT SPARS AS CLOSE TO THE FRONT SPAR AS PRACTICABLE.
7. LOCATE NEW ACCESS HOLES AFT OF THE FRONT SPAR BETWEEN THE FIRST AND SECOND STRINGERS AFT OF THE SPAR. WHEN INSTALLING THE DOUBLER, ROTATE IT SO THE TWO STRAIGHT EDGES ARE CLOSEST TO THE STRINGERS.
8. ALTERNATE BAYS, WITH NEW ACCESS HOLES STAGGERED FORWARD AND AFT OF THE FRONT SPAR, ARE PREFERABLE.
9. A MAXIMUM OF FIVE NEW ACCESS HOLES IN EACH WING IS PERMISSIBLE; IF MORE ARE REQUIRED CONTACT THE SERVICE DEPARTMENT.
10. WHEN A COMPLETE LEADING EDGE SKIN IS BEING REPLACED, THE WING SHOULD BE SUPPORTED IN SUCH A MANNER SO THAT WING ALIGNMENT IS MAINTAINED.

A. ESTABLISH EXACT LOCATION FOR INSPECTION COVER AND INSCRIBE CENTERLINES.
B. DETERMINE POSITION OF DOUBLER ON WING SKIN AND CENTER OVER CENTERLINES. MARK THE TEN RIVET HOLE LOCATIONS AND DRILL TO SIZE SHOWN.
C. CUT OUT ACCESS HOLE, USING DIMENSION SHOWN.
D. FLEX DOUBLER AND INSERT THROUGH ACCESS HOLE, AND RIVET IN PLACE.
E. POSITION COVER AND SECURE, USING SCREWS AS SHOWN.

Fig. 7-11 Access hole installation.

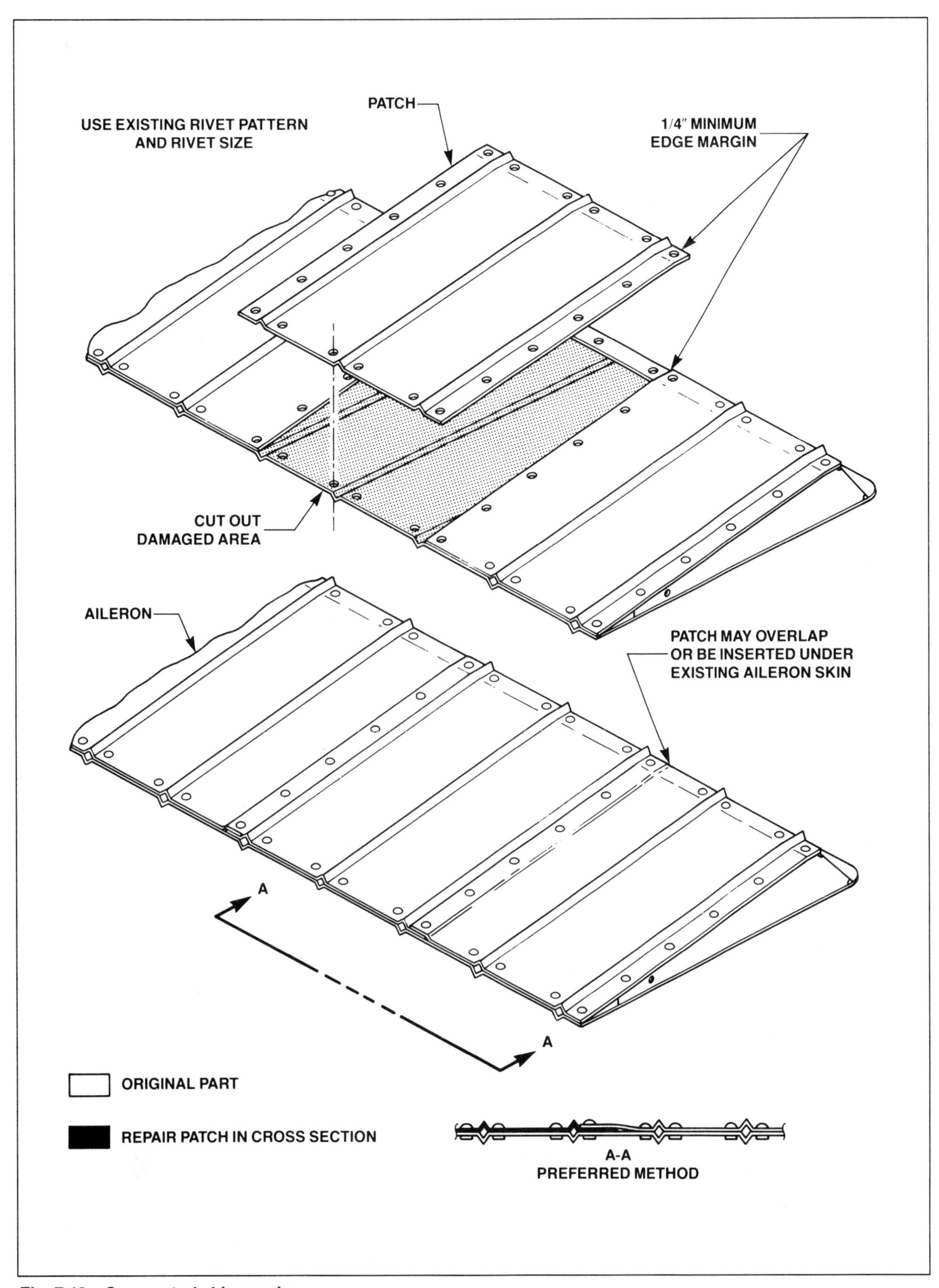

Fig. 7-12 Corrugated skin repair.

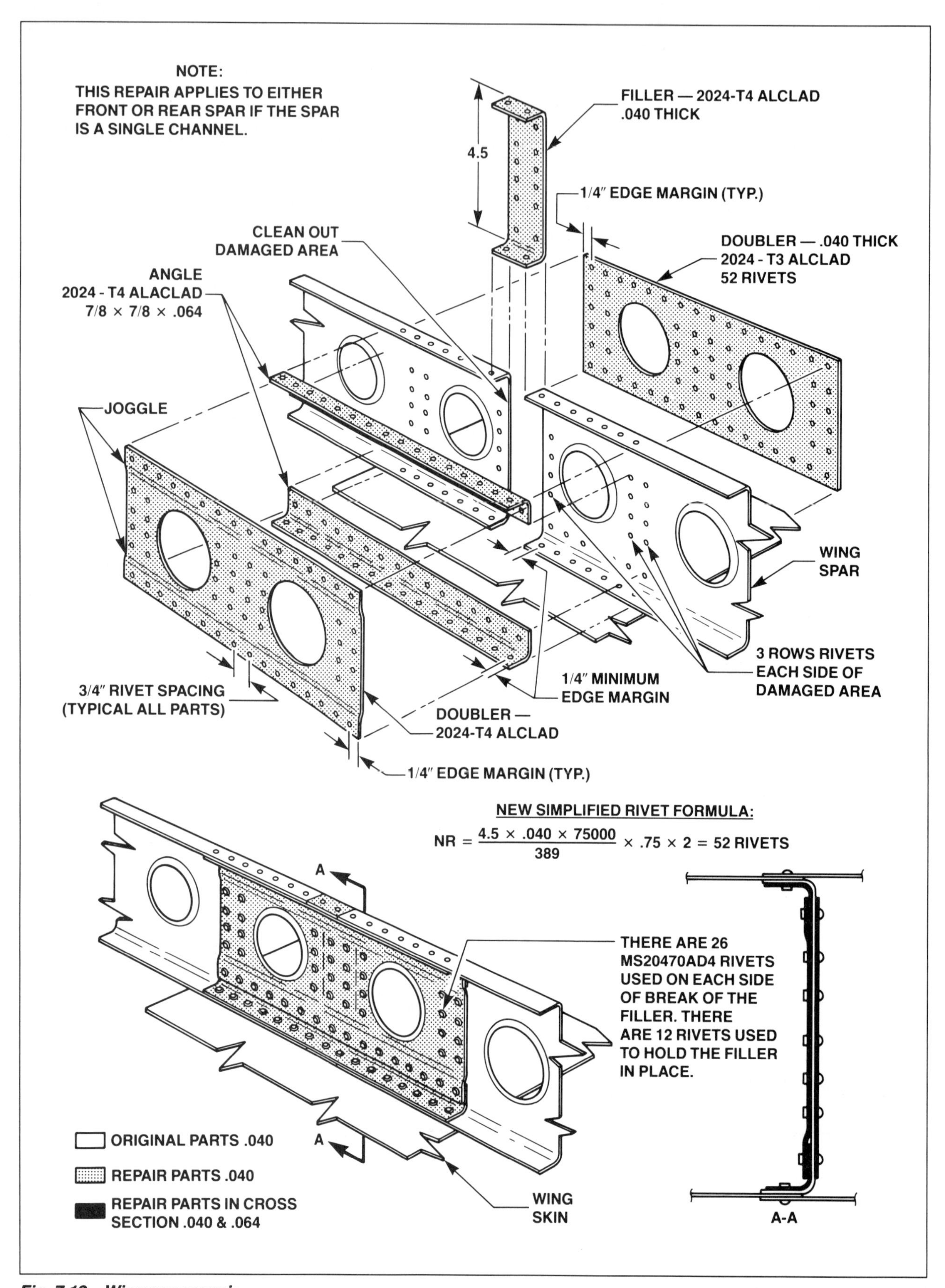

Fig. 7-13 Wing spar repair.

QUESTIONS:

1. *Name three kinds of acceptable data for major repairs.*
2. *List the three major repairs to an aircraft structure.*
3. *What steps should be followed in making a repair or an alteration?*
4. *What is the purpose of the manufacturer's parts manual?*
5. *Name three shapes of patches used on the skins of aircraft.*
6. *What type of patch is used on low-speed aircraft?*
7. *What is the cause of metal cracking?*
8. *Using the rivet formula, find the number of rivets needed for a patch if the break is 4 inches, the metal thickness is .051 of an inch, and AN470AD5 rivets are being used.*
9. *Who signs the 337 form before it is sent to the FAA?*
10. *How many copies of the 337 form must be made?*
11. *Who gets copies of the 337 form?*
12. *What is the formula for finding the number of rivets per inch?*
13. *If the shear strength of a rivet is 459 pounds, what is the total shear strength of 48 such rivets?*
14. *Where are access plates usually located on an aircraft wing?*

Chapter VIII Aircraft Corrosion

A. Surface Corrosion

Surface corrosion, or pitting, is a form of corrosion that eats away at the skin of light aircraft. Surface corrosion usually occurs first in the form of blistering beneath cracked paint.

Surface corrosion can go undetected until it breaks through the metal, when it is too late to save the affected parts. Special attention should be given to the inspection of aircraft which are infrequently flown or are parked in corrosion-prone environments, especially seacoast and industrial areas.

Sulfates and acid base mixtures from industrial smokestacks can penetrate the openings of wings or airfoils. When the heat of the sun causes the wing to sweat, deposits begin to build up in the form of white acid crystals. The first buildup may not be damaging, but in time the metal loses its ability to ward off corrosion. By the time a mechanic detects the corrosion, it may be too late to repair it.

Blistering should be cleaned away with paint remover so that the area can be inspected for the extent of corrosion and then ***alodined*** and repainted. When the corrosive condition is the type that has worked its way through from the inside, the only remedy is skin removal and internal part inspection, a method so costly that thought should be given to replacing the whole part.

Corrosion of the interior surfaces of a wing (Figure 8-1) should be treated as soon as it is detected. The affected parts should be alodined and sprayed with zinc chromate. This "seaplane treatment" can be given to new aircraft.

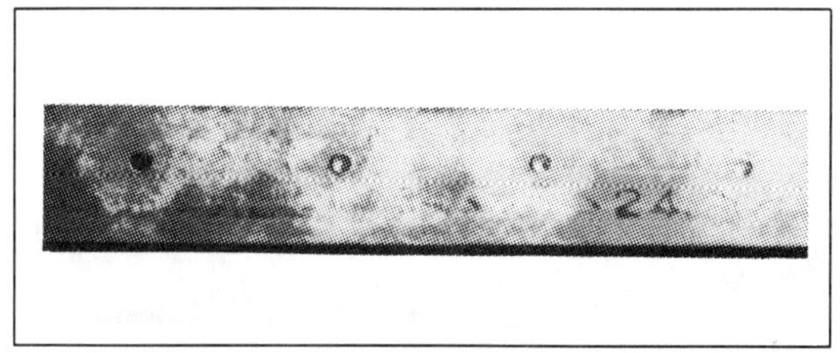

Fig. 8-1 Surface corrosion from the interior. Corrosion attack of stringer located on underside of wheel well. This form of corrosion was caused by moisture being trapped in crevices around wheel well openings.

Other preventative measures are to fly the aircraft frequently in order to air out the insides of the wings, and to protect the aircraft, in a hangar if possible, from elements which accelerate corrosion.

B. Intergranular Corrosion

Intergranular corrosion is caused by improper heat treatment. When aluminum alloys are taken from the heat treating furnace to the quenching tank, intergranular corrosion begins. The elapsed time should not exceed ten to fifteen seconds.

Aluminum alloys such as 2024T3, 2017T3, and 2014T3 are likely to be affected; aluminum alloyed to zinc rarely is. Intergranular corrosion usually does not affect thin aluminum alloys in the 2000 series because they cool rapidly after heat treatment. Thick sheets of copper-bearing aluminum alloys are more susceptible and, when they are to be used on large aircraft, special precautions must be taken during the manufacturing process to prevent intergranular corrosion.

In its early stages, intergranular corrosion can be detected by a microscope.

The problem of intergranular corrosion can be minimized by the use of the aluminum alloy 7075T6, a zinc alloy which is heat treated by the precipitation method.

C. Stress Corrosion

Stress corrosion manifests itself in an intergranular form. It is the result of small cracks which allow oxygen to come in contact with the exposed core metal. The small cracks appear on thick aluminum alloy parts where interference fit holes are drilled. The installed bushing or fastener presses outward and, if the aluminum part is not strong enough, cracks will appear along the lines of force, allowing moisture to enter.

Stress corrosion can best be prevented during the manufacturing process. If it occurs regularly on manufactured parts, an inspection alert or AD note will suggest or specify the type of repair to be made.

D. Dissimilar Metal Corrosion

Dissimilar metal corrosion is a severe and common form of corrosion. It is the result of allowing two metals of greatly differing electronic potential to come in contact with one another in the presence of an electrolyte such as water.

A ***galvanic action*** takes place, similar to that of a lead/acid battery. The active metal behaves like an ***anode*** (+); the less active like a ***cathode*** (−). The transfer of electrons from the more active metal to the less active one causes a very corrosive condition.

When parts made of dissimilar metal must be joined, they should be insulated by non-porous insulators, such as cellophane or thin plastic; dielectric insulators, such as a thin phenolic washer; or at least two coats of zinc chromate primer applied to each contacting surface.

E. Fretting Corrosion

Fretting corrosion results from two metal surfaces rubbing against one another with a very slight motion, such as that caused by ***sympathetic vibration.*** As the two pieces of metal rub against one another, they form small particles which act like ball bearings rubbing against the two surfaces. This continued action causes the surfaces to become harder and harder, until the metal finally cracks.

Early signs of fretting corrosion are trailing, dark deposits behind rivet heads. These dark trails, which look like wisps of smoke or burned oil, are actually fine deposits of aluminum oxide that have worked their way up through the rivet holes.

F. Magnesium Corrosion

Magnesium is the most chemically active metal used in the construction of aircraft. Once magnesium corrosion begins, it proceeds rapidly. It is easy to detect in its early stages because it causes the metal to swell. The corrosion appears as white spots, which quickly develop into mounds which look like white whiskers.

Protection involves the removal of all corrosion products, the restoration of surface coatings by chemical treatment, and a re-application of protective coatings such as paint.

G. Exfoliation

The end result of all corrosion is ***exfoliation.*** It becomes evident in its last stages, when an aluminum part will burst out and shed delaminated layers. Many tiny flakes and scales appear, so that it looks like a stack of needles in shredded wheat.

In this state, the metal no longer has any strength. The only remedy is complete removal of the affected sections. Figure 8-2 shows an example of exfoliation.

H. Corrosion Prevention

1. Alclad

Some sheet metal surfaces of modern aircraft are covered with a coating of pure aluminum called ***alclad.*** Alclad is pressed on to the aluminum alloy part after the metal has been rolled or drawn into its desired shape.

It is approximately 5% of the thickness on each side of the alloyed skin. If the total thickness of the finished skin is .040 of an inch, the alclad covering would be .004 of an inch. Thus, the alloy is .036 of an inch thick. The difference must be considered when replacing non-alclad skins with alclad.

Alclad is used to cover aluminum alloy because the oxides which form on pure aluminum make the surface highly resistant to corrosion.

2. Anodizing

Anodizing is an electrochemical process which causes aluminum oxides to solidify on the surface of aluminum alloys. The oxides are not stationary. They cling and slip around just as a pith ball does on a glass rod loaded with static electricity. The result of the anodizing process is that aluminum oxides are made to stick to the surface of the metal. This action provides a good corrosion-resistant surface and an excellent paint base.

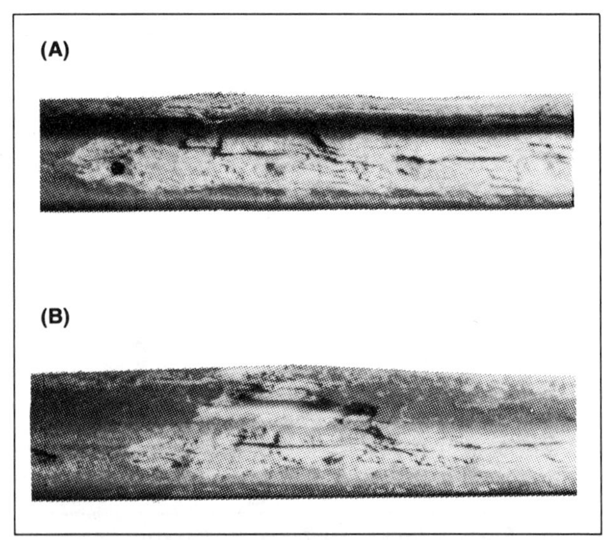

Fig. 8-2A The metal is completely destroyed and only needle-like particles remain.

Fig. 8-2B Notice the swelling at top of metal that is exfoliated.

Anodizing (Figure 8-3) requires a cathode, which is the steel tank that holds the chromic acid; an electrolyte, which is chromic acid; a 40-volt DC supply with an on/off switch; and an anode, which is the aluminum alloy part to be anodized.

The aluminum alloy part is immersed in the chromic acid and the voltage is turned on. The electron flow causes the oxides on the surface of the aluminum alloy to soften. The part is then removed from the tank and dipped into hot water which is mixed with a colored dye. The colored dye is absorbed by the softened oxides. The part is then allowed to dry.

Anodized, blue, hydraulic or fuel system AN fittings are often seen, but anodizing is used for many other exterior parts of aircraft.

3. Alodizing/Alodining

Alodizing is a chemical treatment of aluminum alloy which increases its resistance to corrosion and improves its paint-bonding qualities. It requires no special equipment.

Before alodizing, the aluminum alloy parts should be cleaned with an acidic or alkaline metal cleaner and rinsed with fresh water for ten to fifteen seconds. The alodining chemicals are then applied — by dipping, spraying or brushing — to the surface of the metal.

At this point, a thin, hard coating develops. It will be a slightly iridescent, light bluish-green on copper-free alloys; olive-green on copper-bearing alloys.

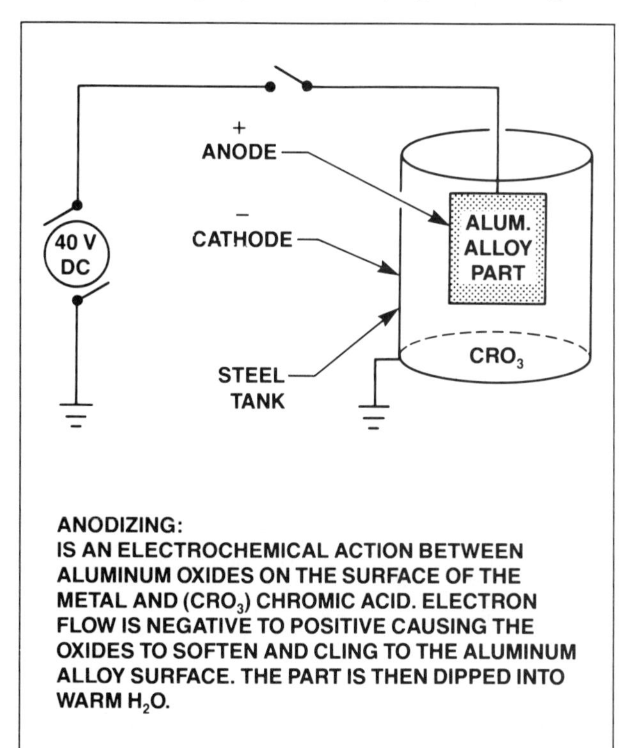

Fig. 8-3 Anodizing.

The alodized parts should be rinsed with clear cold or warm water for fifteen to thirty seconds. Finally, the parts should be given a deoxylyte bath for ten to fifteen seconds. Deoxylyte counteracts alkalinity and makes the alodized parts slightly acidic on drying.

4. Painting Aircraft Exteriors

A good paint finish is an effective barrier between metal surfaces and corrosive environments.

Aluminum alloy surfaces need to have aluminum oxides neutralized before compatible paint primers are applied. If paint is sprayed or brushed directly onto aluminum alloy without chemically treating the oxides, it will peel off almost immediately upon drying. Be especially careful to follow manufacturer's recommendations when preparing the outer surfaces of aluminum alloy aircraft skin to be painted.

Etching is an old-fashioned but successful method of preparing a surface to be painted. Etching is accomplished by washing the outer skin with caustic soda and rinsing with liberal amounts of fresh water. The part is then blown dry with compressed air, and a light coat of zinc chromate is applied to the whole surface.

All aircraft should be periodically stripped of paint down to the bare metal and inspected for corrosion before a new paint job is begun. It is highly recommended that any metal aircraft paint job be preceded by an alodine treatment.

I. Cleaning Aircraft

When the aircraft is cleaned, it should be closely examined for signs of corrosion, such as paint flaking and blistering. Ordinarily, paint flaking is localized and can be removed by a chemical paint remover. If no further corrosion is found, alodine the exposed area and repaint.

QUESTIONS:

1. *What form of corrosion is caused by improper heat treatment of the alloy?*
2. *What is the name of the corrosive action that occurs when two sheets of metal rub together and form stress cracks?*
3. *What type of corrosion results from cracks caused by pressed-in bushings?*
4. *What type of corrosion causes the inner surfaces of the metal to begin to blister through to the outside?*
5. *Name a process used to protect the skins of aircraft.*
6. *What does anodizing do to the surface of aluminum alloy?*
7. *What must be done to the surface of aluminum alloy before any primer or paint is applied to it?*
8. *What is the name of the process used by aluminum manufacturers to protect the surface of aluminum alloys?*
9. *What happens to an aluminum alloy part affected by intergranular corrosion?*
10. *Which of the following is the most corrosive: a) aluminum, b) magnesium, or c) steel?*

Chapter IX
Applied Forming, Bending And Layout Of Aircraft Parts

In modern aviation there are not many sheet metal mechanics who can still make aircraft parts like they did years ago. In the 1940s and '50s about the only way aircraft parts could be obtained was to have them made by a sheet metal mechanic who knew how to shape, bend and form metal.

As the aviation industry grew, the demand for new parts increased. Aircraft manufacturers began making parts which were approved and inexpensive. The shop mechanic no longer bothered to make parts as he once did, and much of the expertise of the trade was lost.

An area in the aviation community where the old skills of forming and bending are having a rebirth is in the aircraft "home builder" industry. These aircraft builders are learning many of the skills of the past and are doing a good job making their own parts. The instructional materials showing how to form, bend and assemble parts are not readily available.

The main purpose of this chapter is to show the application of layout, bending, forming and assembly of aircraft parts. The soaring costs of aircraft over the last several years is another reason for showing examples of how to construct aircraft parts. These high prices have forced many fixed base operators and flight schools into completely overhauling and rebuilding their aircraft rather than replacing them as they did in the past.

The rebuilding consists of replacing the interiors, stripping the paint from the skins, replacing old or worn out parts and overhauling the engines. New parts may not be available for some of the older aircraft, so they will have to be made by the sheet metal mechanic as they once were years ago.

Many aircraft home builders do not have the professional training of an A&P mechanic, consequently they get into trouble by not using the correct aircraft structural materials. An understanding of aluminum alloy and the different aircraft approved steels is essential for the safe construction of a home-built aircraft.

Chapter III contains the various metals and alloys used for the construction of the aircraft and its parts. A review of the metals used on aircraft should be examined before learning how to develop, lay out, form and construct its parts. It has already been established, in this book, that modern civilian aircraft are made primarily from aluminum alloy. Therefore the examples used in this chapter will deal with forming, bending, layout and assembly of aluminum alloys.

The internal and external structural parts of an aircraft always require the use of heat treated aluminum. The most popular type of alloy used for skins, spars, stringers and longerons is A2024-T3 or A2024-T4. The difference between the two is the T3 is heat treated, age hardened and coldworked, while the T4 is only heat treated and age hardened.

There are parts, such as inspection covers, wheel caps, wing tips and prop spinners where non-heat-treatable aluminum or annealed 2024-0 is used. The kind of steel recommended for the construction of an aircraft is 4130 chrome-molybdenum. Whenever there is any doubt, always consult the aircraft manufacturer, an A&P mechanic knowledgeable about sheet metal, or an FAA authorized inspector.

A selection of aircraft parts used to show the various techniques of construction are: a joggled joint, a metal tube, a nose rib, a layout and assembly of a control surface, and the construction and installation of an access plate with an inspection cover. The construction of these parts will stress one or more techniques of planning, developing (proper material selection), layout (bend allowance and setback), hand bending, brake bending, forming or assembling.

A. Construction Of A Joggled Joint

A joggled joint is made to allow for an offset of a piece of sheet metal so it will fit flush with a stringer, rib or spar stackup, as shown in Figure 9-1.

A ***joggle*** is sometimes used at the trailing edges of wings or control surfaces to provide a finished trailing edge and a flat surface for the installation of rivets. Figure 9-2 shows a joggle used on light aircraft which increases joint strength and reduces weight by eliminating the need for stringers or former rings.

Joggles are also used to form an offset with large extrusions so the flow of the metal will not be interrupted. Whenever a piece of metal is joggled, its strength is increased due to coldworking.

There are two ways to joggle an aircraft part: The manufacturer's method, where a special die is used; or the shop method, where hand or floor tools are used. The methods used in the shop vary according to the mechanic's needs as the aircraft is being built or repaired.

One of the most common methods used to make a joggle is by forming with a bending brake. The steps of brake joggling are shown in Figure 9-3.

When joggles are needed on narrow strips of metal (e.g. with a width of four inches) a hard wood block can be cut on a jigsaw to the desired shape and forced into shape using C-clamps, as shown in Figure 9-4.

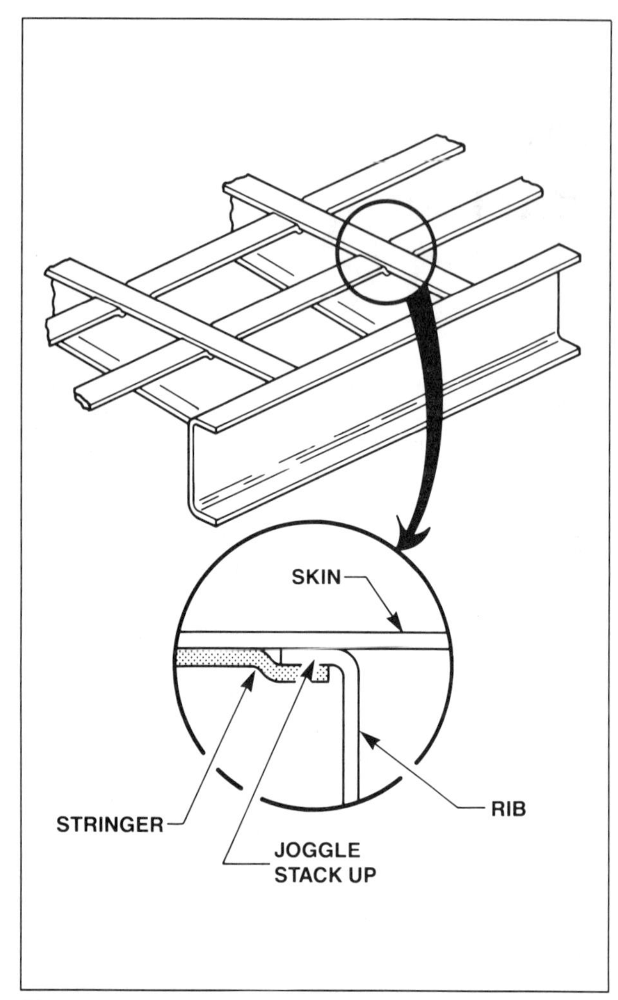

Fig. 9-1 A stackup of a joggle joint consists of a rib, joggle stringer and skin.

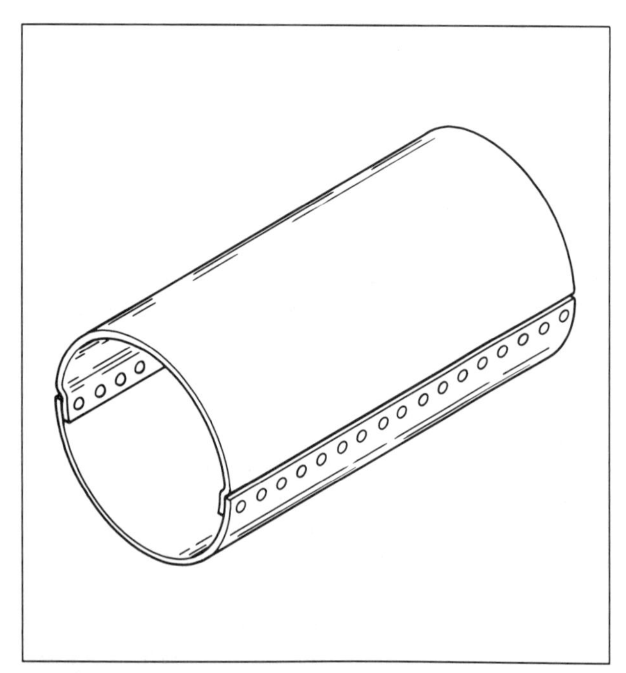

Fig. 9-2 Joggle skins replace stringer in fuselage.

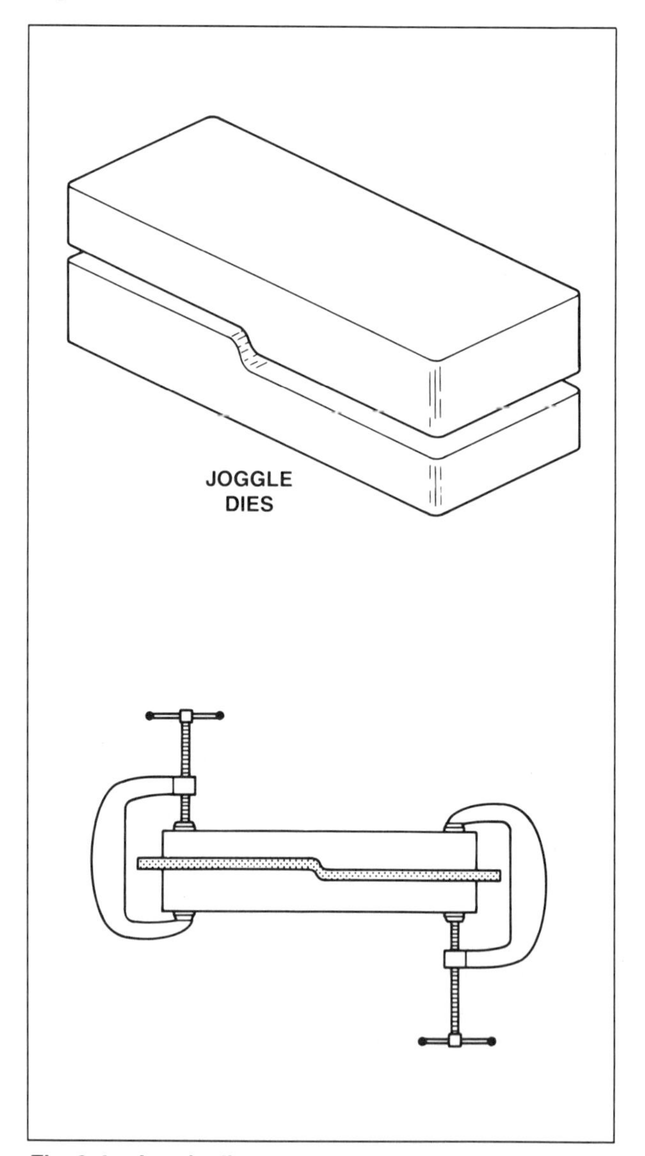

Fig. 9-4 Joggle dies.

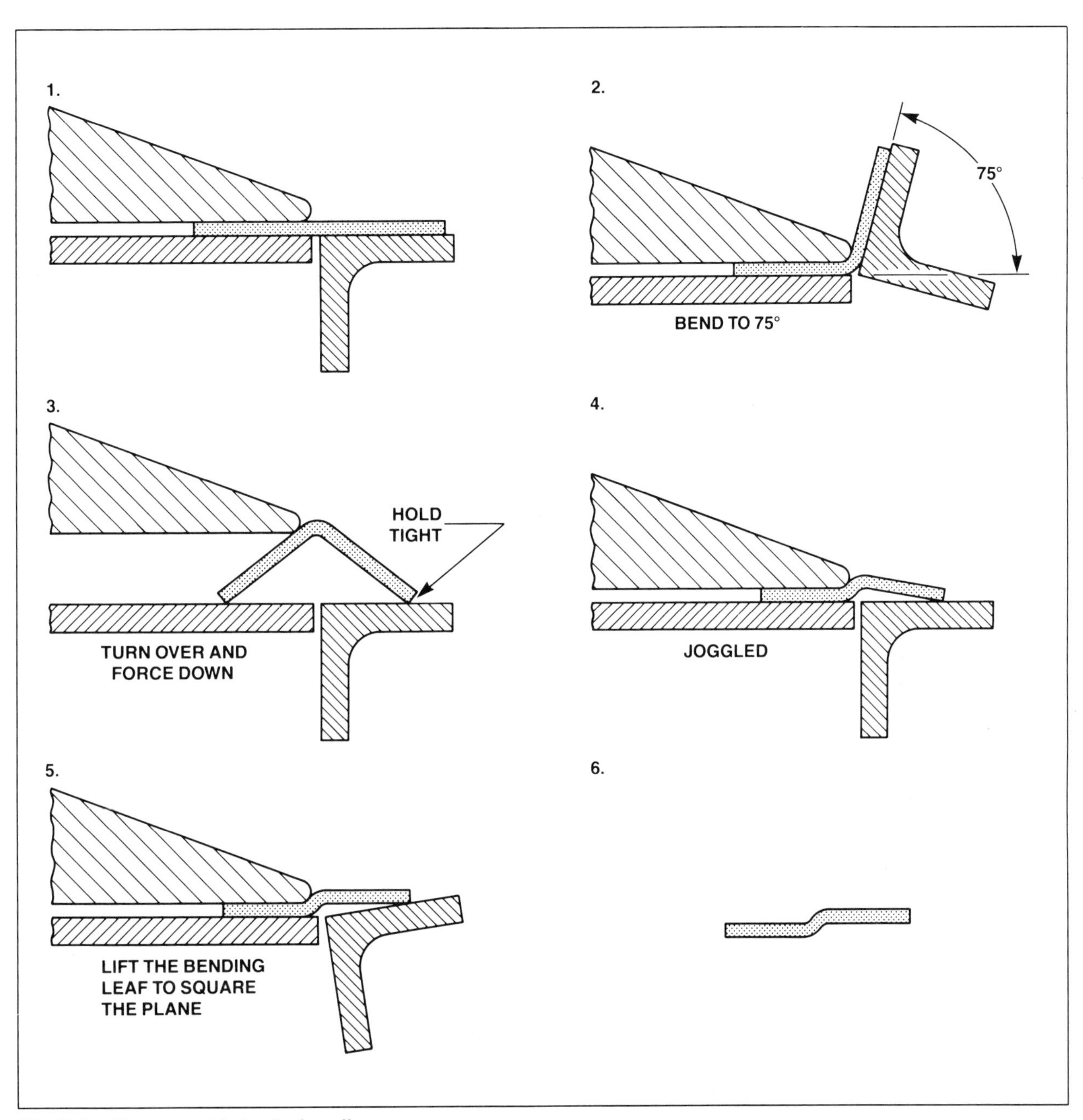

Fig. 9-3 Steps used in brake joggling.

Materials List

1. 18 × 12 sheet of A2024T3 aluminum alloy
2. Felt marking pen
3. Pencil, layout paper and eraser

Tools List

1. Six inch rule
2. Straight edge
3. Box-pan brake or bending brake
4. Bench shears or hand shears

The first part of this operation requires knowing the amount of joggle needed. The trailing edge of a control surface skin 18 inches long needs to be joggled.

Mark off a line one inch in from the trailing edge. Place the metal onto the bending brake using the line just drawn as a sight line. Bend the metal to approximately 75 degrees. Turn the metal over and raise the movable bending brake jaw to its wide open position. Place the metal under the brake as shown in Figure 9-3.

Hold the metal so it will not slip and pull down on the bending brake handle. This action will force the metal to joggle. After the arm is forced down the two metal planes will not be in alignment, so pull up slightly on the folding leaf of the bending brake.

B. Construction Of Aluminum Tube

A large diameter tubing can be used for ducting hot or cold air to various places in the aircraft. There are two types of tubing used for this purpose: One is flexible and the other is rigid.

On modern aircraft, flexible aluminum tubing is not used as often any more because it is being replaced with specially treated fabric covered ducting. The use of rigid tubing still serves a valuable purpose because it can take more abuse than soft flexible ducting.

Materials List

1. 18 × 8 inch sheet of 5052-0 .050 thick aluminum alloy
2. Pencil, drawing paper and eraser
3. 20 each AN470A3-6 rivets, or
4. 20 each AN426A3-6 rivets
5. 2 × 4 board three feet long

Tools List

1. Bench or hand shears
2. Slip roll former
3. Ball peen hammer
4. Number 40 drill
5. Drill motor
6. Double seaming stake or 1½″ diameter pipe
7. Vise with pipe jaw

Either type, heat treated or non-heat-treated aluminum alloy in the annealed condition, is best suited for the making of rigid tubing.

The tool used to make tubing or duct pipe is a set of ***slip rollers.*** The smallest diameter that can be made using a slip roller is determined by the size of the rolling machine. When a tube diameter larger than the roller is called for, the rollers can be adjusted to make a larger diameter such as those needed for leading edge construction.

The construction of an aluminum alloy tube requires knowing the diameter, length and type of locking joint to be used. Suppose that a 3-inch diameter tube will be made. A finished aluminum tube with a 3-inch diameter in an isometric view, with the formula for finding the circumference of a circle is computed for your information in Figure 9-5.

The circumference of the tube is the length of metal needed to roll the diameter of the tube. With a diameter of 3 inches times 3.1416, the length of roll out is 9.42 inches, and the finished tube length is 18 inches. To each of the 9.42 inch ends add one inch to make the joggle and locking seam as shown in Figure 9-6.

Once the joggle and seam are bent, release the slip roller and lock the flat sheet into place in the middle as shown in Figure 9-7. After the slip roller is locked into place, adjust the rollers to begin the initial curvature of the tube. Crank the rollers back and forth and roll the tube from one end to the other until the tube begins to take shape.

Each time you roll, gradually increase the tightness of the rollers until the tube is completely rolled into shape. Release the slip roller and remove the tube from the roller. Check the diameter and seams for alignment. Once the seam is joined, slip it over a pipe clamped in a bench vise as shown in Figure 9-8 and begin flattening the seam. This action will lock the seam into place.

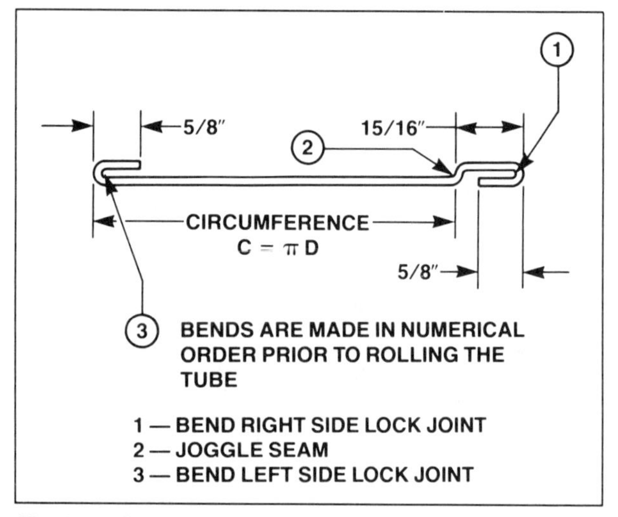

Fig. 9-5 Construction formula for aluminum alloy tube.

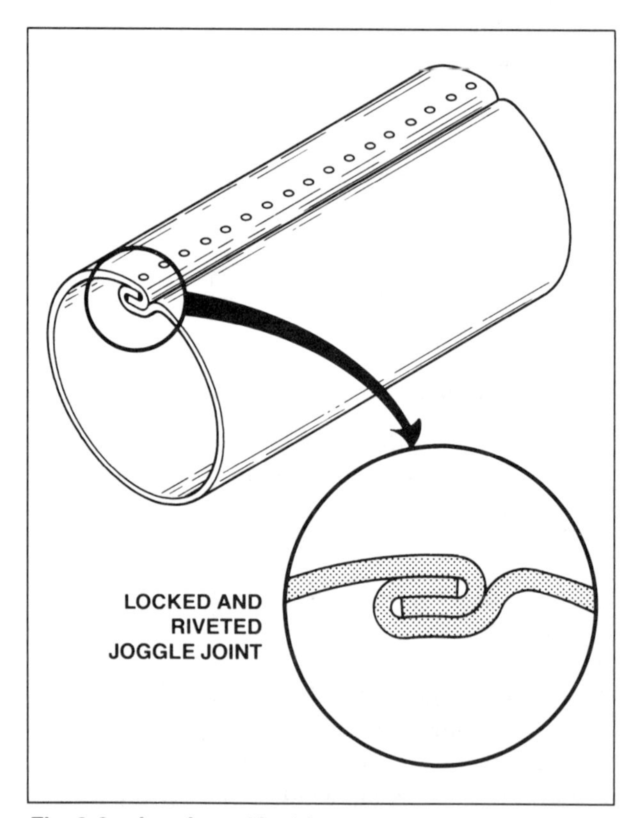

Fig. 9-6 Joggle and locking seam.

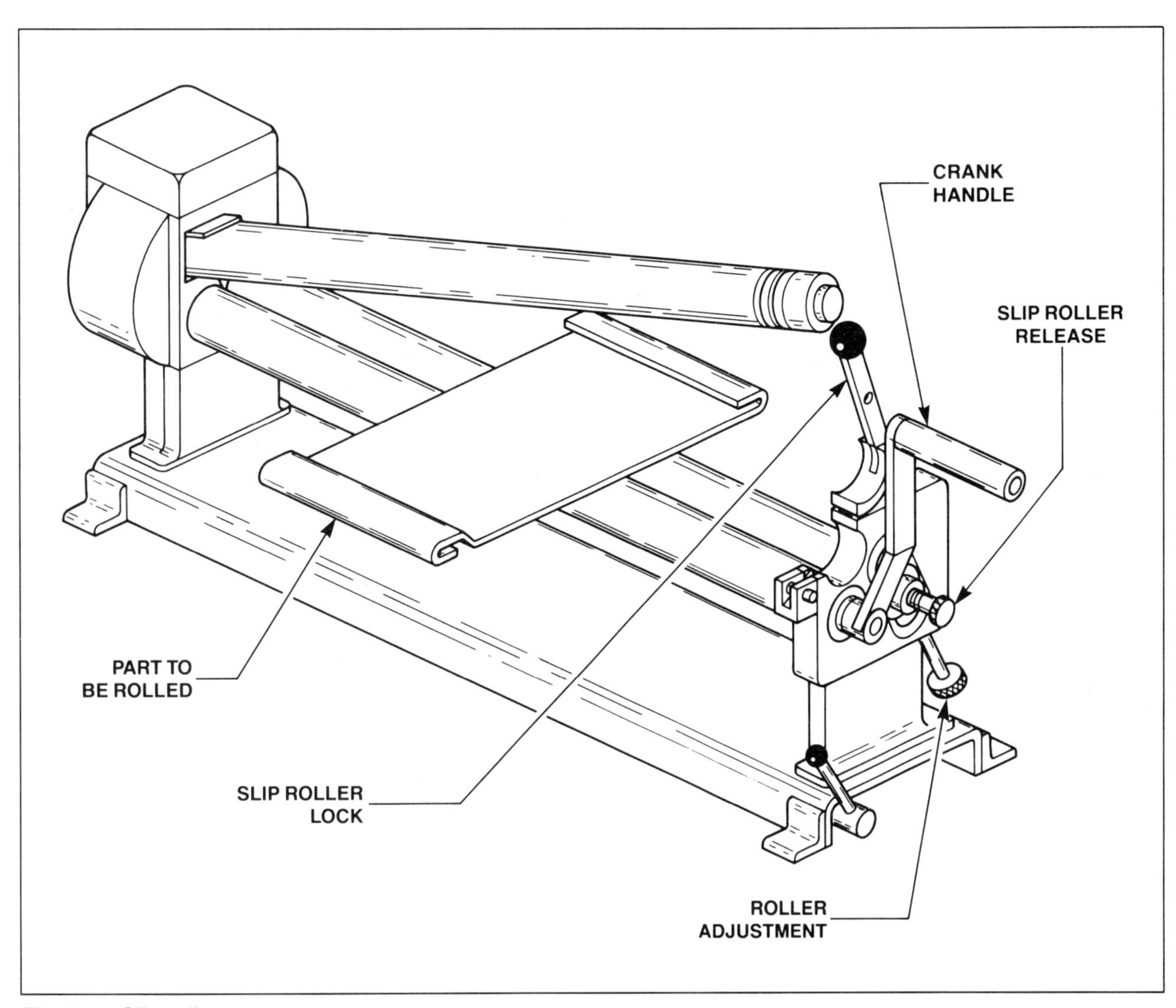

Fig. 9-7 Slip rollers.

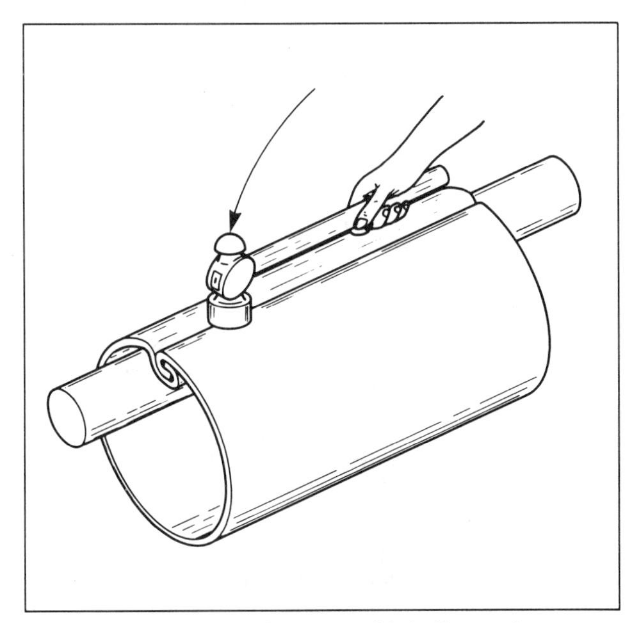

Fig. 9-8 Flatten joggle seam with ball peen hammer.

If more joint security is needed you may want to install soft aluminum rivets (AN470A3-6) along the seam. The holes for the rivets are drilled by slipping the tube over a pre-formed 2 × 4 clamped in a bench vise. The rivets can be bucked using the same pipe you used to lock the seam. If this piece of tubing is going to be fitted into another piece of tubing, the rivets used to secure the seam must be countersunk (AN426A3-4).

C. Construction Of A Nose Rib

Nose ribs are also called former ribs and are not considered primary structure. The nose rib consists of a flange, lightening hole and a rib-to-spar flange. The term flange can be interchangeable with the term ***cap strip.***

Nose ribs are riveted to the front of the spar which connects to a former or compression rib on a wing or control surface. The installation of a nose rib completes the airfoil shape of a wing leading edge or control surface.

The type of metal most often used to make a nose rib is A2024T4. The A2024T4 aluminum alloy is not coldworked and therefore is easier to form over a smaller radius. In the area of the bend the metal becomes coldworked, making it stronger.

A steel die or hard wood block is used to form a wing or control surface nose rib. Some examples of the various designs of nose ribs are shown in Figure 9-9. The nose rib we will be using for our example is the type with notches and relief holes.

Materials List

1. Hardwood block
2. A2024T4 aluminum sheet stock
3. Sand paper, 400 grit
4. #6 wood screws
5. Pencil, drawing paper and eraser

Tools List

1. Jigsaw, hand or machine
2. Belt sander
3. Wood rasp
4. Hole saw
5. Drill motor
6. #30 drill
7. #21 drill
8. Mallet
9. Hand shears or snips
10. Center punch
11. Ball peen hammer
12. Pair of hand benders

1. Development Of The Nose Rib

To develop a nose rib, the length, height, thickness, radius and number of degrees the metal will be bent must be known. These measurements can be found in the manufacturer's prints.

The nose rib made from A2024T4 .025 thick has a finished size of 4 1/4 inches long and 4 inches high. The flanges are 5/8 of an inch high with a bend radius of 1/8 of an inch.

Draw the full size outline of the rib on drawing paper, showing the extension of the flanges from the bend allowance area. The flange used to connect the rib to the spar is 5/8 of an inch. Calculate bend allowance and setback using a metal thickness of .025, a bend radius of .125 and a bend angle of 90 degrees.

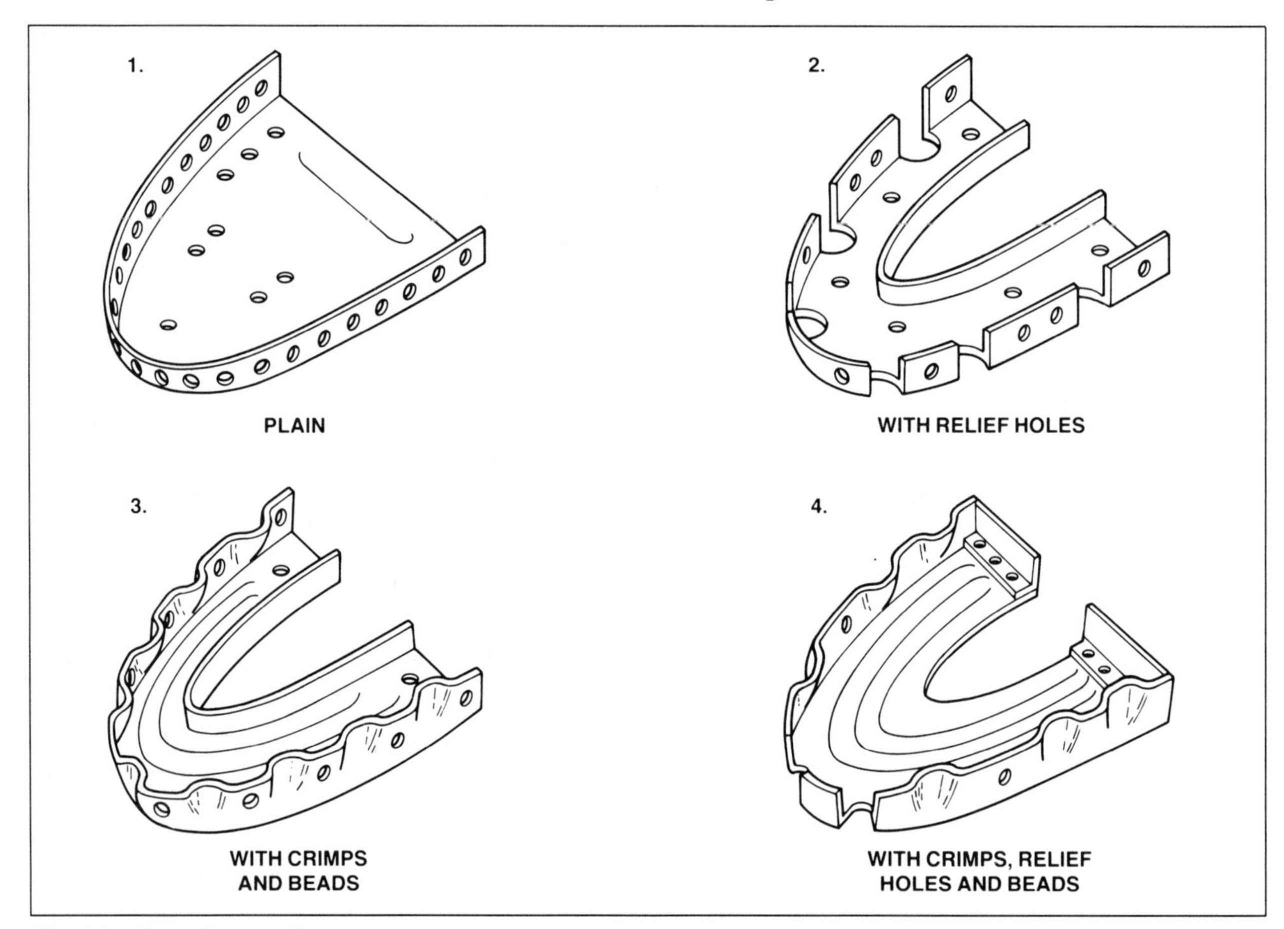

Fig. 9-9 Sample nose ribs.

(A)

ITEM	FINISHED	UNBENT LEG LENGTHS	FRACTION
A	4.000	3.700	3-45/64
BA		.214	7/32
B	4.250	3.950	3-61/64
BA		.214	7/32
C	.625	.475	15/32
T	—	—	—

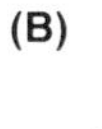

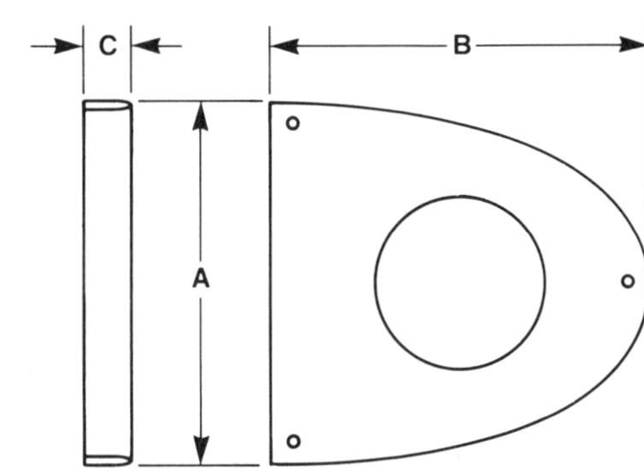

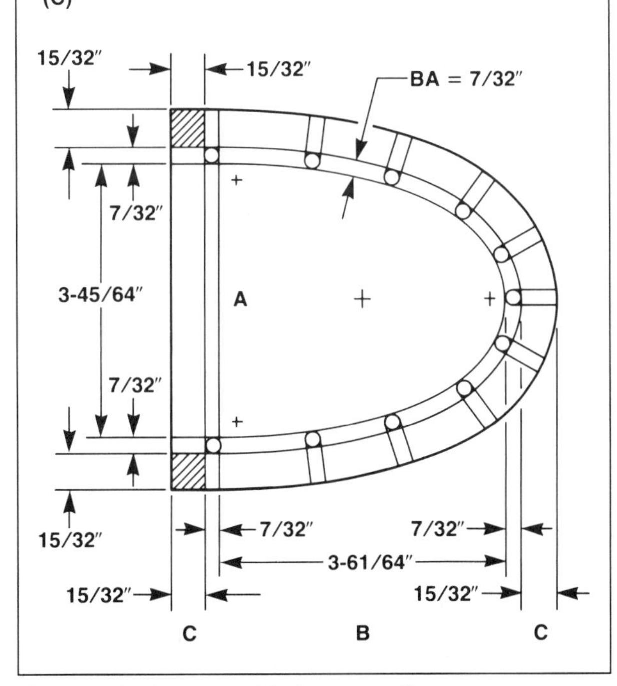

Fig. 9-10A Layout table for nose rib.
Fig. 9-10B Finished nose rib.
Fig. 9-10C Layout lengths.

The value for bend allowance can be found in the bend allowance table (Appendix A-II), or by using the empirical formula from Chapter VI. You will find that bend allowance is .214. Setback for a 90 degree bend obtained by adding R .125 and T .025 is .150. Figure 9-10 shows a layout table containing the rib leg lengths and bend allowance.

2. Layout Of The Nose Rib Block

The ***forming block*** used to make this nose rib is made from hard wood. When constructing a nose rib, the proper shape of the rib is very important, therefore the calculations must be as accurate as possible. The measurements are taken from a manufacturer's print or drawing, and a pattern is made on a sheet of drawing paper. The pattern is then transferred to the wood block.

The wooden block can be sawed, using either a hand or machine-operated jigsaw. It is important to leave a little extra material around the saw line so the rib can be brought to size using a belt sander. The edges of the rib block must be rounded to a bend radius of .125 of an inch. The sides of the block where the flanges are formed should be shaped slightly more than 90 degrees to allow for spring-back. All block edges must be shaped by a belt sander or a hand wood rasp, and finish smoothing with 400 grit sand paper.

The lightening hole is cut out using a rotary hole saw as shown in Figure 9-11. The lightening hole

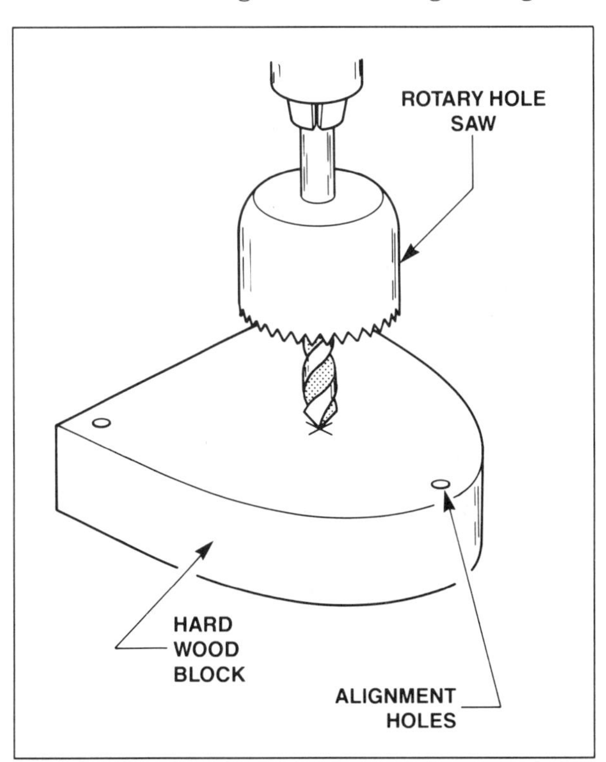

Fig. 9-11 Rotary hole saw used to cut lightening hole.

(similar to a giant countersink) is beveled 30 degrees using a half round file to shape the angle. After making the bevel, a hard wood dowel is cut to a countersink tapered flange of 30 degrees to form the bevel around the lightening hole. The combination of the forming block and the dowel is shown in Figure 9-12.

3. Layout And Cutting The Nose Rib

Measure and mark the layout lengths onto the rib stock metal. Set off segments of 3/4 of an inch around the nose of the rib. At the point where the lines touch the line of sight in the bend allowance area, mark it with a center punch. Use a number 21 drill to make relief holes at each 3/4 of an inch intersections around the nose of the rib.

After drilling out all the relief holes use a pair of straight hand shears and cut slots equal to the diameter of the #21 drill hole. At the spar tab end of the rib, drill a relief hole at each intersection and cut the corners using a pair of straight cut hand shears.

Next, mark the center of the rib and cut the lightening hole to size, using a Greenlee hole punch. Before going on to the next step, cut enough stock to make all the ribs needed in this size. Trace the layout of the rib onto each of the rib stock sheets, mark all bend lines, forming marks, relief holes and hold down screw holes.

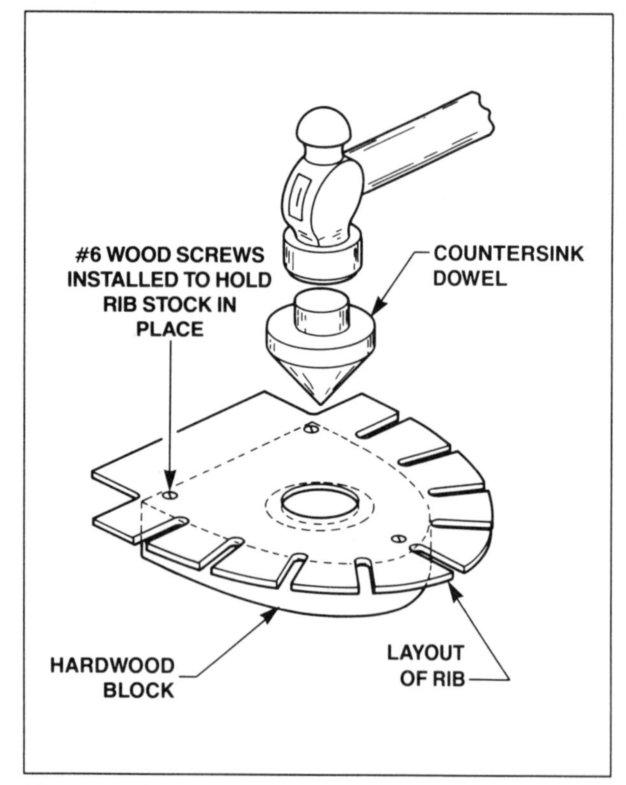

Fig. 9-12 Forming block and dowel.

4. Forming The Nose Rib

Before placing the rib stock over the forming block, drill three #30 holes through the rib and into the block, as shown in Figure 9-12. Insert the three #6 wood screws into the holes. They are used to hold the rib stock in place while bending the flange and forming the lightening hole.

Begin forming the rib by bending the spar tab first, then hammer the two forward tabs of the rib flange, as in Figure 9-13. Continue forming the remaining flange material.

With the rib secured by the three wood screws, begin forming the countersunk beveled flange around the lightening hole. The countersunk flange is formed by pounding the tapered dowel rod into the beveled portion of the forming rib. Remove the three wood screws and examine the rib for squareness. If necessary, trim the end of the flanges so the rib will be even all the way around the nose.

D. Layout And Construction Of A Control Surface

The construction of a control surface or wing is basically the same with respect to the development, layout and assembly of its three main parts: the ribs, spar and skins.

A mechanic must be familiar with the mathematical formulas and know how to work with the bend allowance table to successfully plan a layout of a control surface. Information on how to use the mathematical formulas for bend allowance and setback are covered in Chapter VI. A table for bend allowance values is found in Appendix A-II.

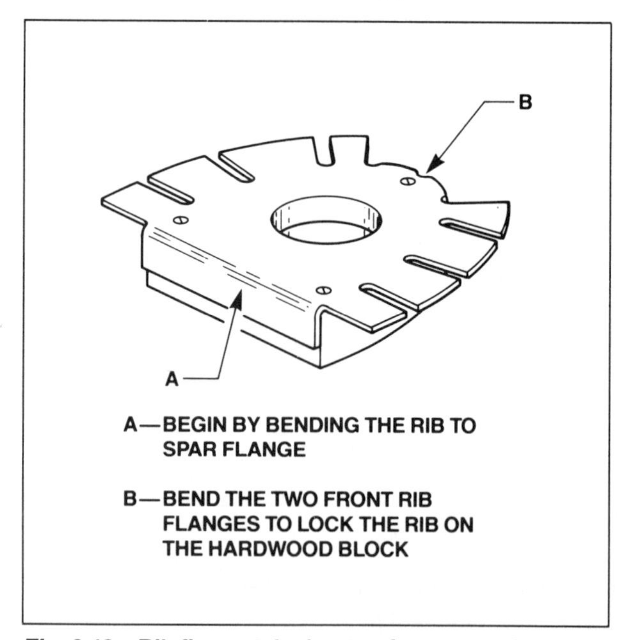

Fig. 9-13 Rib flange tabs bent to form nose rib.

Materials List

1. One 18 × 12.5 sheet A2024T3 .032 thick (skin)
2. One 18 × 12.5 sheet A2024T3 .040 thick (skin)
3. Three 5.25 × 12 sheets A2024T3 .025 thick (ribs)
4. One 18 × 6 sheet of A2024T3 .040 thick (spar)
5. Pencil, drawing paper and eraser
6. AN470AD3-6 rivets

Tools List

1. Box-pan bending brake
2. Bench shears
3. Straight edge
4. Six inch scale
5. 4 each C-clamps
6. Drill motor
7. Half round file
8. Number 40 drill
9. Silver #40 clecos
10. Three pound bucking bar
11. Countersink gun set
12. Universal head gun set
13. Countersink cutter
14. Hand benders

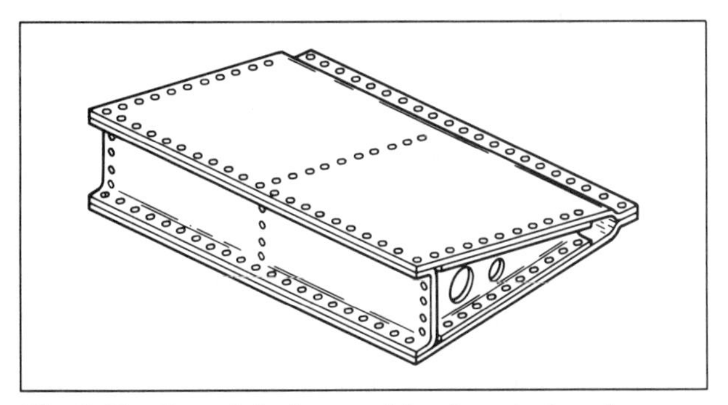

Fig. 9-14 Completed assembly of control surface.

Prior to constructing a control surface, its size and shape, type and thickness of metal and number of degrees of bend for each part must be known. This control surface will show the applications of bend allowance, setback, layout, forming and assembly.

This control surface is 18 inches long. Its inboard and outboard ends are twelve inches wide. For simplicity, and because the parts and assembly are repetitious, we will show the construction of the first three inboard ribs, spar and skins. Understand, the layout and construction of this control surface could just as well be seven feet or longer if needed.

The type of material used for the construction of this control surface will be A2024T3 aluminum alloy. The ribs are .025 thick, the spar is .040 thick, and skins are .032 and .040 thick. The rivet stock is AN470AD3-5 which will be used on all the joints and skins. The finished length of the ribs is 10 1/8 inches. The spar end of the rib is 3.5 inches and the trailing edge is 3/4 of an inch. The flange that mates to the spar at the front of the rib is 3.5 inches before it is cut to size and the spar cap is 3/4 inches long. The skins are 18 inches long by 12.5 inches wide. The trailing edge of the control surface skin will be joggled to increase strength and provide a flat surface for the installation of the rivets. Figure 9-14 shows a completed assembly of a control surface.

1. Development And Layout Of Ribs

The first parts to be made are the ribs. These ribs connect the front spar and trail off into a joggled trailing edge.

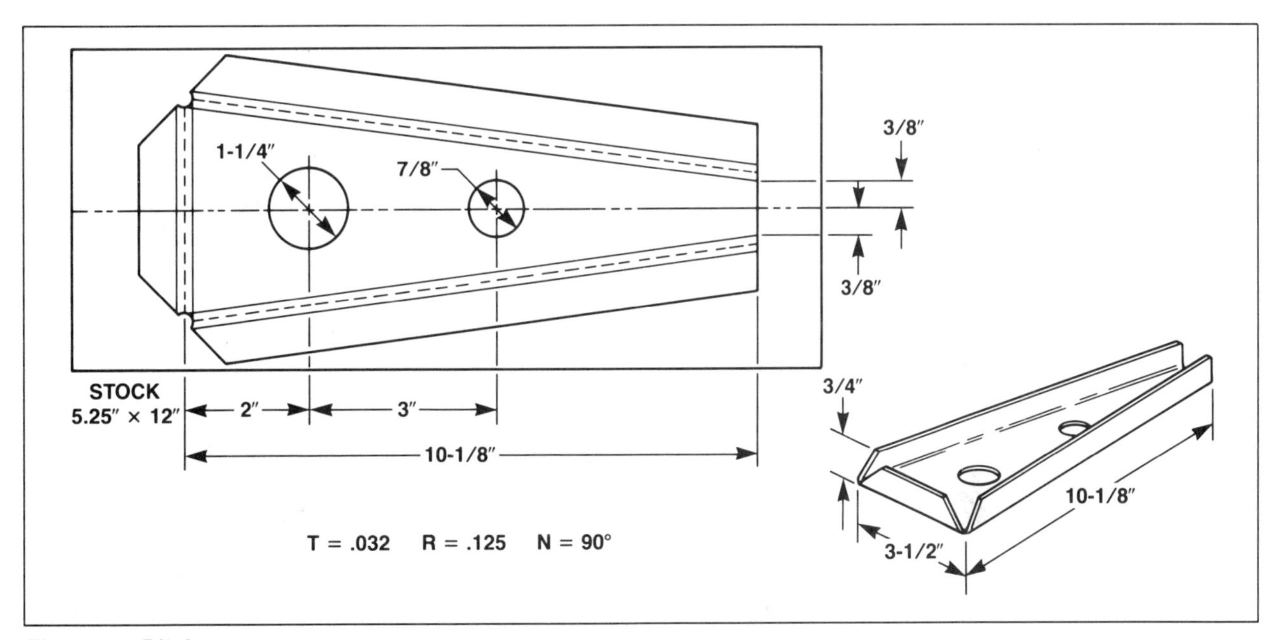

Fig. 9-15 Rib layout.

The ribs are made from .025 thick A2024T3 aluminum alloy. Bend allowance is computed by using a bend radius of .125 of an inch, metal thickness of .025 of an inch, and 90 degree flange angles.

Bend allowance taken from the table in Appendix A is .214 of an inch. Setback R = .125 plus T = .025, or .150. Remember, setback is subtracted from finished leg lengths to find the unbent portion so it can be laid out accurately on the rib sheet metal stock.

The size of sheet metal on which the ribs will be laid out is 5.25 × 12 inches. Figure 9-15 shows the layout details of the rib. The length of the rib is determined by measuring 10 1/8 inches along the centerline from the front tab line of sight. The rib-to-spar and main flanges are 3/4 of an inch wide. The layout data shown in Figure 9-17 are taken from the table shown in Figure 9-16.

2. Bending And Forming The Ribs

The best tool to use for marking the layout measurements is a sharp felt marking pen. Begin by plotting the rib centerline on the 5.25 × 12 sheet of aluminum. Split the layout distance of the center (3 13/64 inch) leg into 1 19/32 of an inch on each side of the centerline.

Add to each side the bend allowance and flange leg lengths. Measure 10 1/8 inches along the rib centerline and draw a parallel line 3/8-inch either side of the centerline. From this line draw a parallel line at a distance equal to the bend allowance, and another parallel line at a distance equal to the leg or flange length as shown in Figure 9-17.

After all lines are clearly marked, lay down the lines of sight. The line of sight is one bend radius out from the bend line which is located under the nose of the brake.

ITEM	FINISHED LENGTH	UNBENT LEG LENGTH AND BA	FRACTION
A	.750	.600	39/64
BA		.214	7/32
B	3.500	3.200	3-13/64
BA		.214	7/32
C	.750	.600	39/64
T	—	—	—

R = .215 T = .025 N° = 90

Fig. 9-16 Rib layout chart.

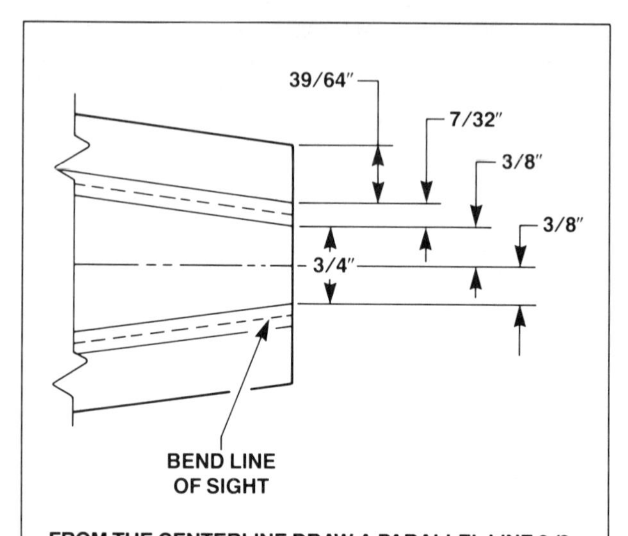

FROM THE CENTERLINE DRAW A PARALLEL LINE 3/8-INCH EITHER SIDE OF THE CENTER LINE. FROM THIS LINE DRAW A PARALLEL LINE AT A DISTANCE EQUAL TO THE BEND ALLOWANCE, AND ANOTHER PARALLEL LINE AT A DISTANCE EQUAL TO THE LEG OR FLANGE LENGTH. THE RESULT IS THE TRAILING RIB HEIGHT.

Fig. 9-17 Trailing edge of control surface rib section.

Begin by bending the main rib flange first. Bend the flange which connects to the spar in the same manner as you would for bending a pan. Before bending the first rib, duplicate its layout on the remaining rib stock. Trace the cutout rib, then place the location of the lines of sight, mark the lightening holes and cut the relief tabs at the spar connect tab.

After bending all the ribs be sure to check their squareness against one another and the against the height of the spar.

3. Development And Layout Of The Spar

After the ribs are formed, they must be measured accurately before the spar can be made. The finished spar is made from .040 thick A2024T3 aluminum to a size of 3/4 × 3.5 × 3/4 inches. The finished rib should be 3.5 inches high, and the height of the spar must be exactly the same. The spar layout and associated chart data are shown in Figure 9-18.

The spar caps may need to be spread open after they are made to conform to the slope of the ribs as shown in Figure 9-19.

4. Bending And Forming The Spar

Begin the forming by bending either one of the spar caps first. Line up the nose of the brake with the line of sight and bend the spar caps to 90 degrees. Repeat this action for the other spar cap. Immediately check the height of the spar with that of the ribs.

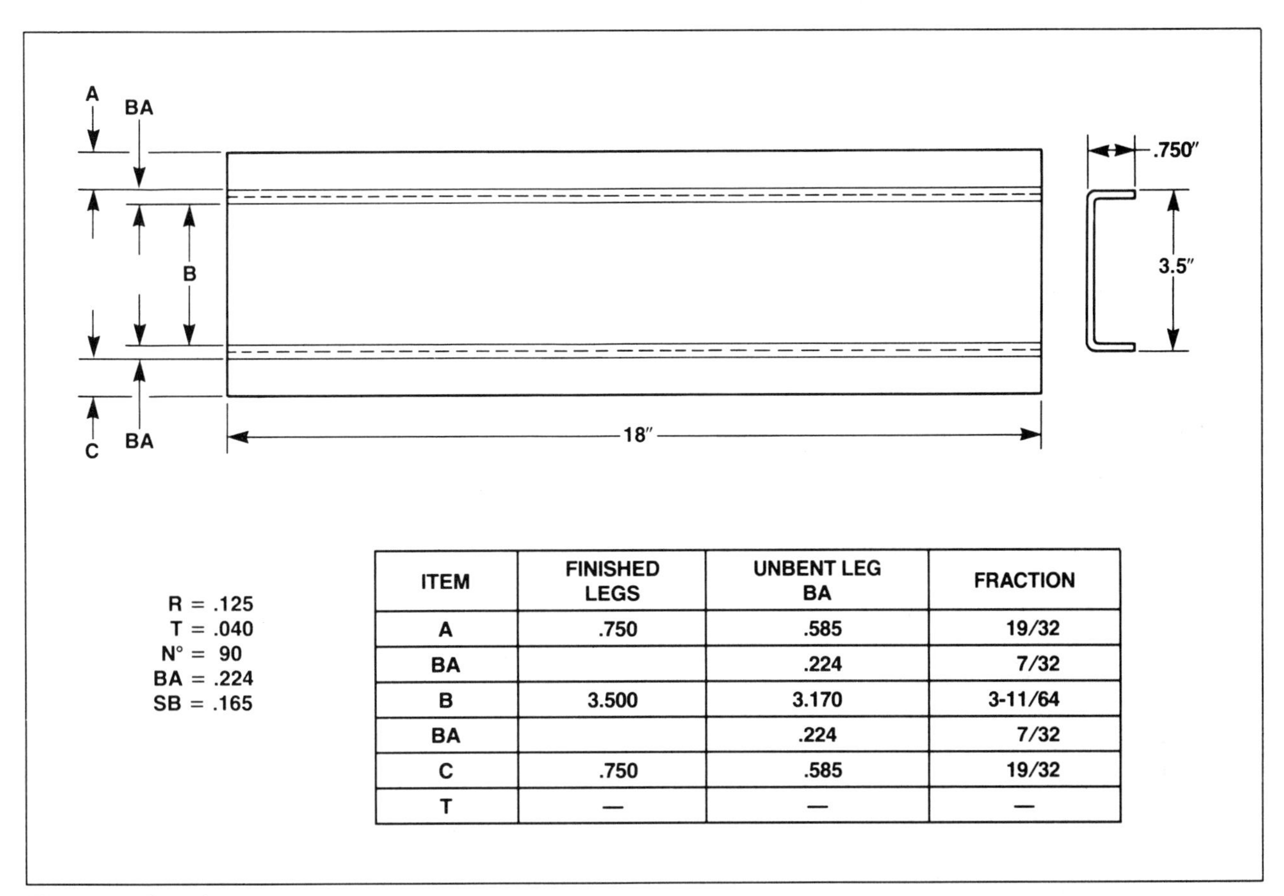

ITEM	FINISHED LEGS	UNBENT LEG BA	FRACTION
A	.750	.585	19/32
BA		.224	7/32
B	3.500	3.170	3-11/64
BA		.224	7/32
C	.750	.585	19/32
T	—	—	—

Fig. 9-18 Spar layout.

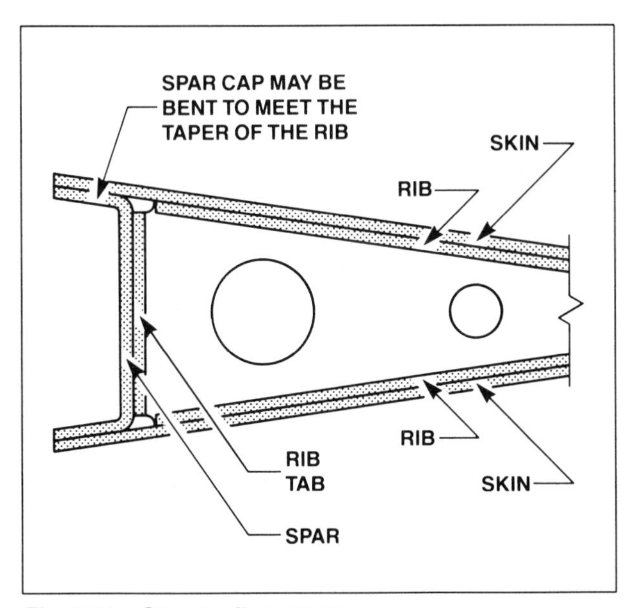

Fig. 9-19 Spar to rib mate-up.

5. Development And Layout Of Skins

Two A2024T3 aluminum alloy skins are used to cover the exterior of the control surface — one is .032, the other .040 thick. The .032 skin requires AN470AD3-6 rivets; the .040 skin uses AN426AD3-6 rivets. To finish off the trailing edge the skins will need to be joggled. The layout of the skins and the rivet pattern to be used are shown in Figure 9-20.

6. Bending And Forming Skins

The most difficult part of forming the skin is the proper layout and bending of the trailing edge joggle. Figure 9-21 shows a detailed view of the method of estimating the amount of joggle needed. The joggle must be deep enough to make the transition from the ribs to the trailing edge as smooth as possible. Another reason for the joggle is to provide a flat surface for the trailing edge rivet installation.

Begin the skin shaping process by clamping the skins on the clecoed assembly of the ribs to spar. With the skin flat against the spar and ribs, check the slope of the trailing edge. Gauge the amount of joggle and divide it between the two skins. The joggles will be made using the bending brake technique as shown in Figure 9-3. After the joggles are made, once again clamp the skins into position and check for squareness.

7. Assembly And Construction Of The Control Surface

While the clecos and C-clamps are still in place, use a felt marking pen and make line up marks on

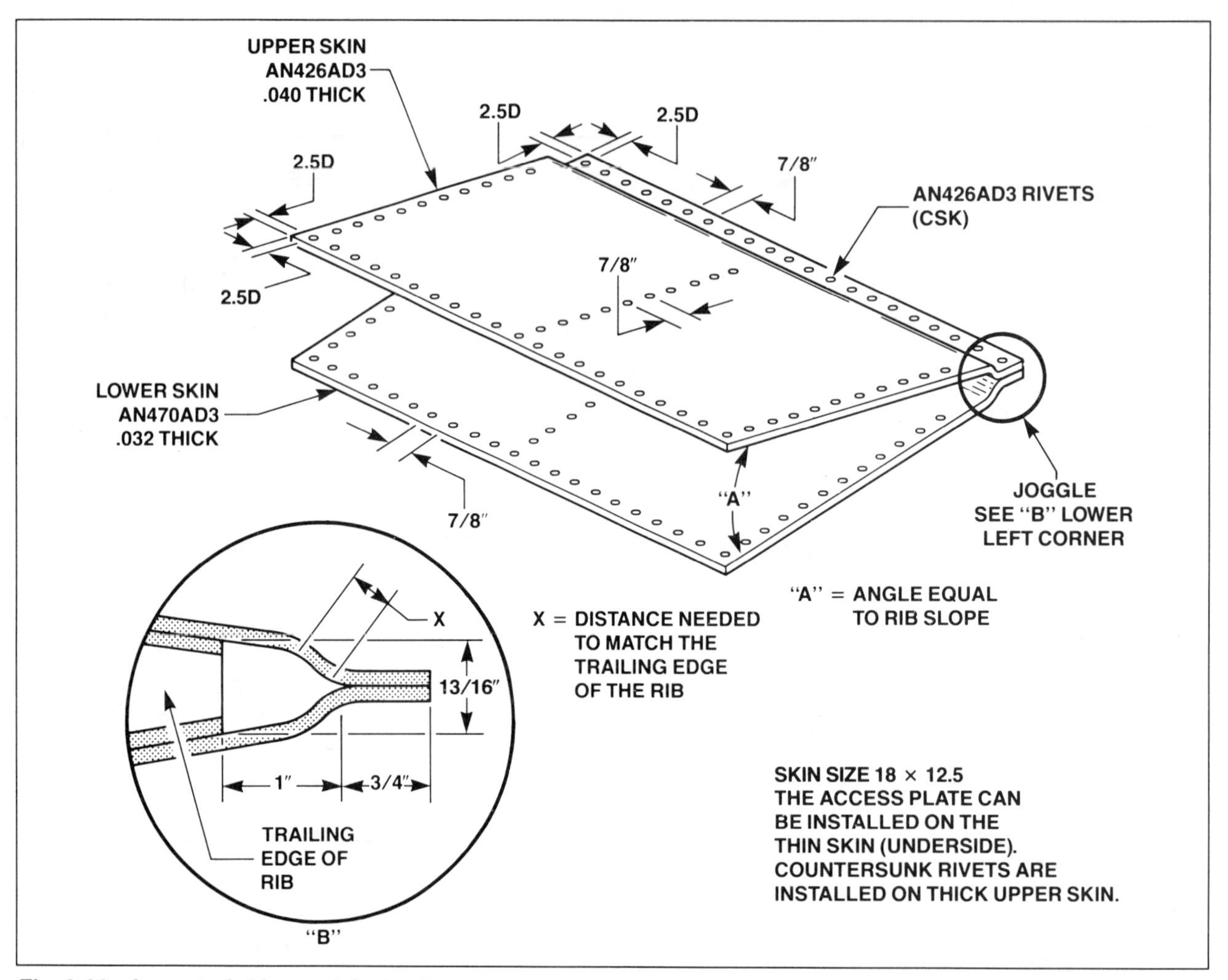

Fig. 9-20 Layout of skins and rivet pattern.

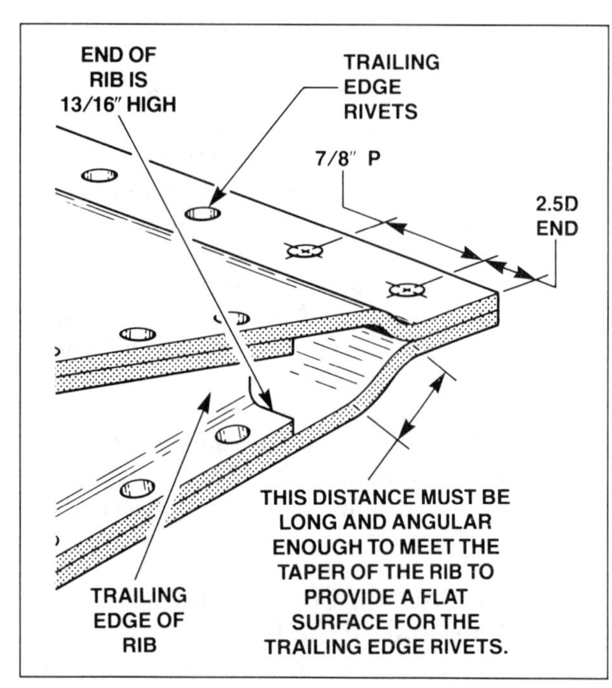

Fig. 9-21 Joggle trailing edge.

all mating surfaces. The reason for placing the line up marks is reassembly will be much easier after drilling and deburring.

Start the final assembly by drilling out all the rivet holes first. The center rib is the first to be drilled as well as riveted. The center rib is riveted to the spar and then to both skins. The first skin can be installed with little or no interference, but the placement of the other skin will require the need for a long thin bucking bar to install the rivets.

After the center rib is installed, the skins can be riveted to the spar. The next operation is to clamp the outer ribs into place and drill out the rivet pattern shown in Figure 9-20.

The next rib to be installed is the one that faces opposite from the center rib flanges. This rib will be installed with its flanges pointing outward. The remaining rib will have its flanges facing in the same direction as the center rib. Both end ribs are referred to as finishing ribs because their flanges are facing outward, making it easy to install the rivets.

After the ribs are installed, then the trailing edge is riveted. Upon completion, be sure to check the control surface for squareness. File off all rough edges and give all rivets a very close inspection.

E. Construction And Installation Of Access Plate And Cover

Access plates are installed on metal aircraft so parts such as electrical wires, fluid fittings and control cables located internally can be inspected. Access holes and plates are usually located on the under side of a wing, along the sides of a fuselage or on the undersides of control surfaces.

Some manufacturers of light aircraft limit the number of access holes to five for each wing. When more than five access holes are needed per wing, always consult the aircraft manufacturer before installation.

The construction of any access hole doubler requires that it be strong enough to make up the loss of strength created by the removal of metal when making the hole. The location of an access hole should be as near as possible to a rib spar or stringer to borrow from its strength.

Materials List

1. Pencil, drawing paper and eraser
2. 7 × 7 sheet of A2024T3 .040 thick (doubler)
3. 7 × 7 sheet of A2024T3 .032 thick (cover)
4. Felt marking pen
5. Masking tape
6. 6 each 8-32 machine screws
7. 6 each 8-32 plate nuts
8. Zinc chromate primer
9. AN426AD3-6 rivets

Tools List

1. Right or left hand shears
2. Compass
3. Protractor
4. Drill motor
5. Silver (3/32 inch) clecos
6. Rivet gun
7. Bucking bar
8. Half round file
9. Rotary file
10. Screwdriver
11. Hand nibblers
12. Countersink cutter
13. Countersink gun set
14. #40 drill

Access plates can be made economically in the shop by following the manufacturers' suggested plans outlined in their maintenance manuals. The plans will give the necessary information pertaining to hole size limits, thickness of metal, required sized rivets and how to space the rivets and cover plate hold down fasteners.

1. Layout Of Access Plate

The proper selection of materials is one of the most important first steps when making an access hole. If the skin of the aircraft is A2024T3 .025 thick, then the access doubler will be A2024T3 .040 thick. The increased thickness of the metal gives an increase in strength, serves as a stiffener and permits (when necessary) the driving of larger diameter rivets into what would normally be thinner skins.

The access plate is drawn to full size on drawing paper. The cover should be the same thickness as the skin. Using a protractor, lay out a pattern for the rivets and the six plate nuts as shown in Figure 9-22. Later the pattern can be transferred to a piece of sheet metal as shown in Figure 9-23. At the same time, the layout of the cover plate can be done on a similar piece of sheet metal.

2. Layout And Forming Steps Of Access Plate and Cover

When making an access hole, the first step is to mark the exact location of the inspection hole. Drill a 3/8-inch starting hole in the aircraft skin. After drilling the hole in the aircraft skin, use a pair of either right or left hand cutting shears to cut a hole about 1/32 smaller then the size of the required hole. The last 1/32-inch is left so the hole can be smoothly dressed to size using a half round file.

The second step is to lay out the doubler ring on a sheet of metal by transferring its measurements from the drawing. Cut out the doubler ring using a Greenlee hole punch, hand shears or nibblers to exactly the same diameter as the hole made in the aircraft wing skin. The doubler is wide enough to allow for the installation of the rivets and inspection plate fasteners as shown in Figure 9-24.

Cut the outside circle of the doubler to size using either left or right hand cutting shears. After the holes are drilled into the doubler, use it as a guide to drill the holes into the aircraft skin. File all edges. Be sure there are no burrs or sharp edges remaining.

The third step is the construction of the cover plate: Fit the doubler over the cover plate material and trace its outline. Cut out the cover plate with hand shears to the same size as the outside diameter of the doubler. Once the circular cover plate is made, fit it under the doubler as shown in Figure 9-25, tape it in place and center punch the machine screw holes. Use the doubler as a guide and drill the holes using a #11 drill. Mark the

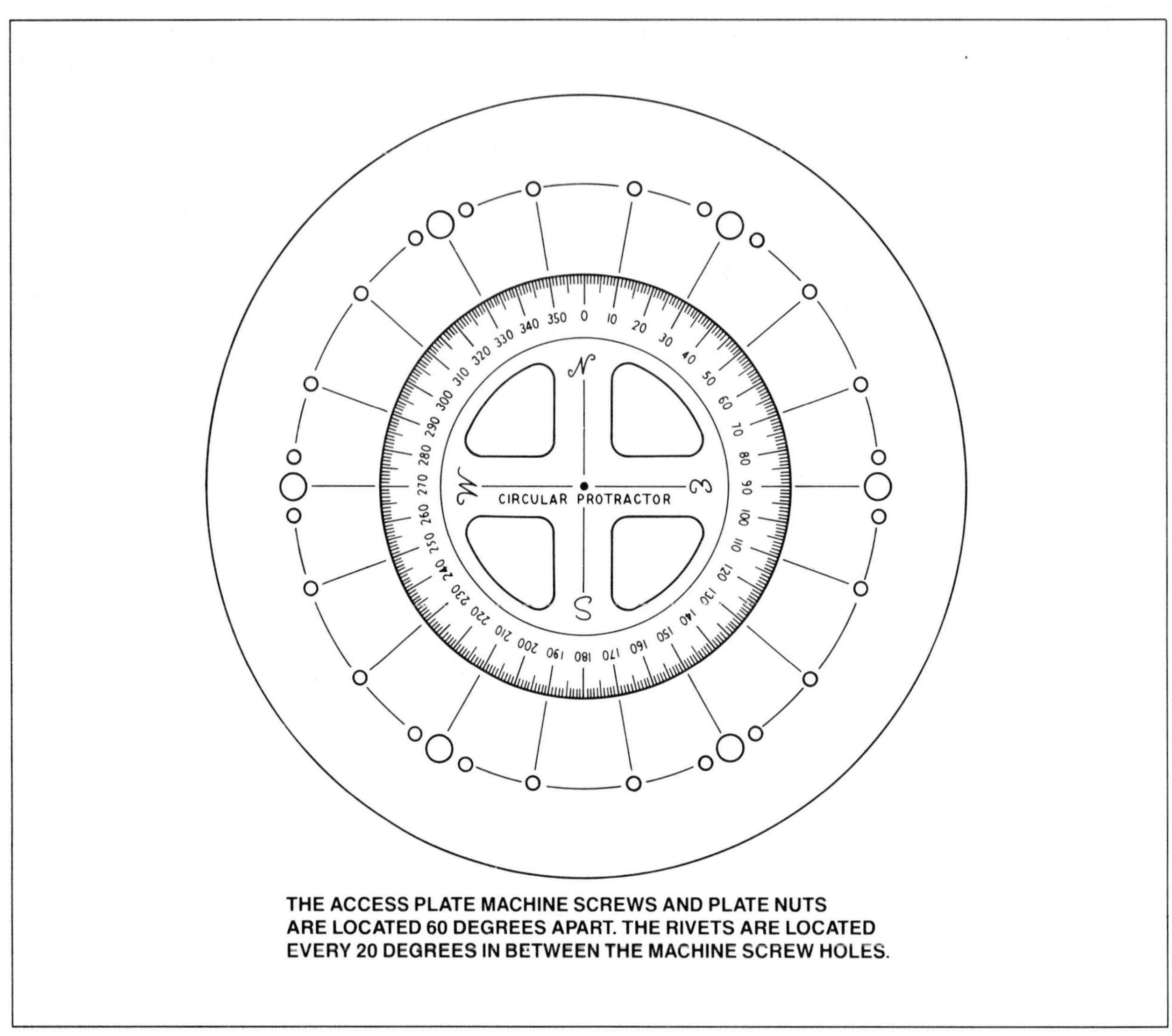

Fig 9-22 Layout of rivet pattern for doubler and skin. Inside diameter is 5", outside diameter 7".

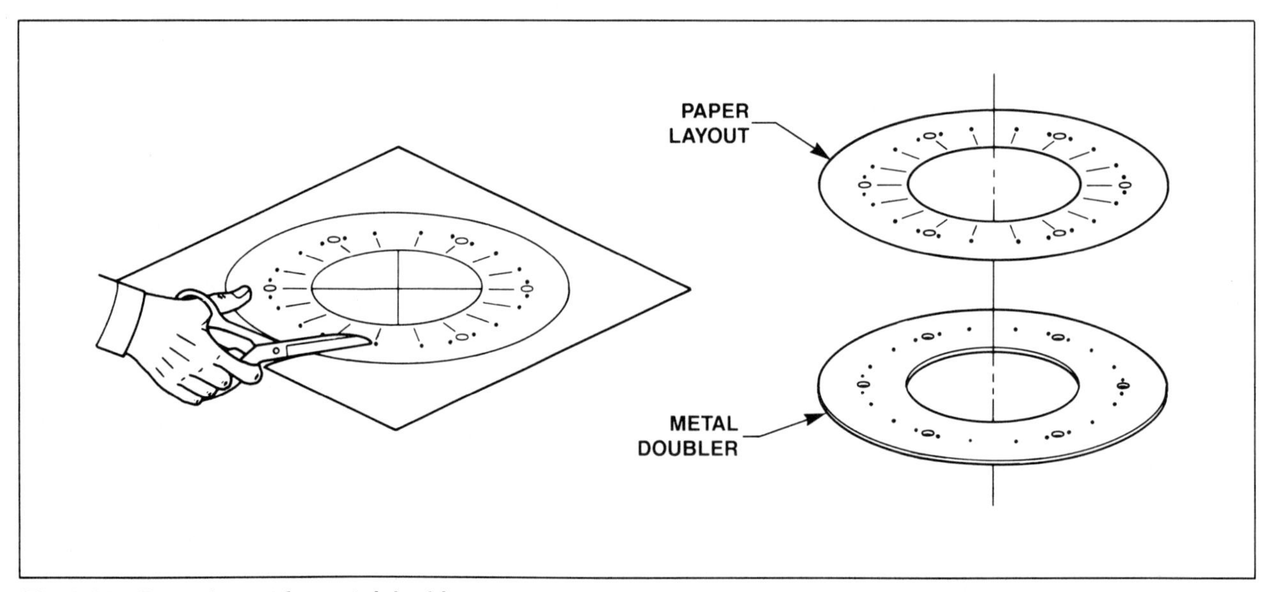

Fig. 9-23 Paper layout for metal doubler.

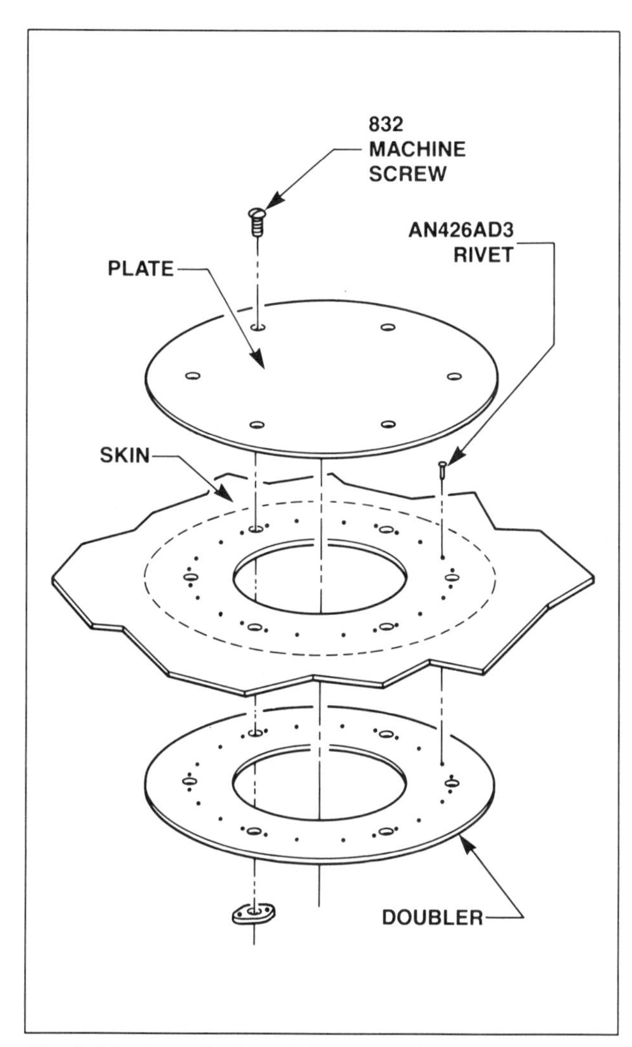

Fig. 9-24 Installation of rivets and inspection plate fastener.

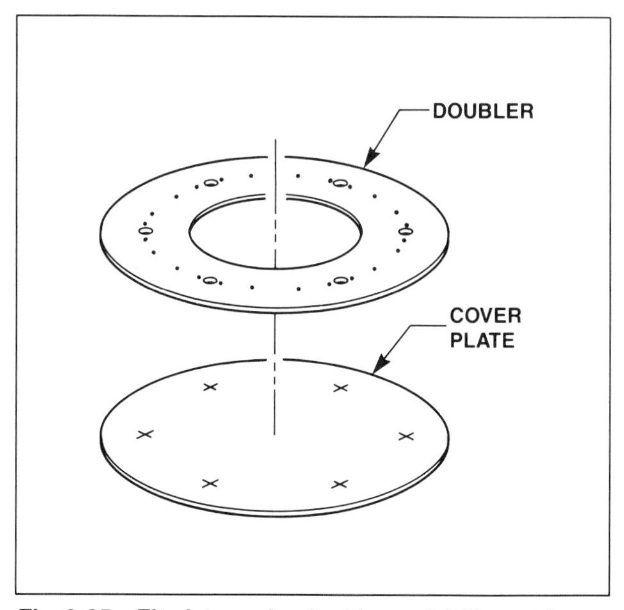

Fig. 9-25 Fit plate under doubler and drill machine screw holes.

relationship of the doubler to the skin and the inspection plate so it will line up properly after you remove the burrs from its holes.

3. Installation Of The Doubler Plate

After cutting the doubler as shown in Figure 9-26, fit it into place and use clecos to hold in place. Drive two rivets 180 degrees apart and fit the cover over the doubler being sure that all six machine screw holes line up exactly before continuing to rivet the doubler in place.

After all rivets are driven, screw the inspection plate in place using six machine screws. If the inspection plate does not lay flat, bend the edges slightly as shown in Figure 9-27.

F. Conclusion

The examples shown in this chapter were selected because they include many of the operations usually involved in the manufacturing of aircraft parts. The making of aircraft parts is time consuming but economical, especially for the aircraft home builder. Making aircraft parts is sometimes the only way to get the parts needed to build or repair an aircraft.

The most expensive item when making aircraft parts is the labor. If you are an aircraft home builder, that is a fact you must well understand from the beginning. As you get involved with the making of aircraft parts, you will begin to learn shortcuts.

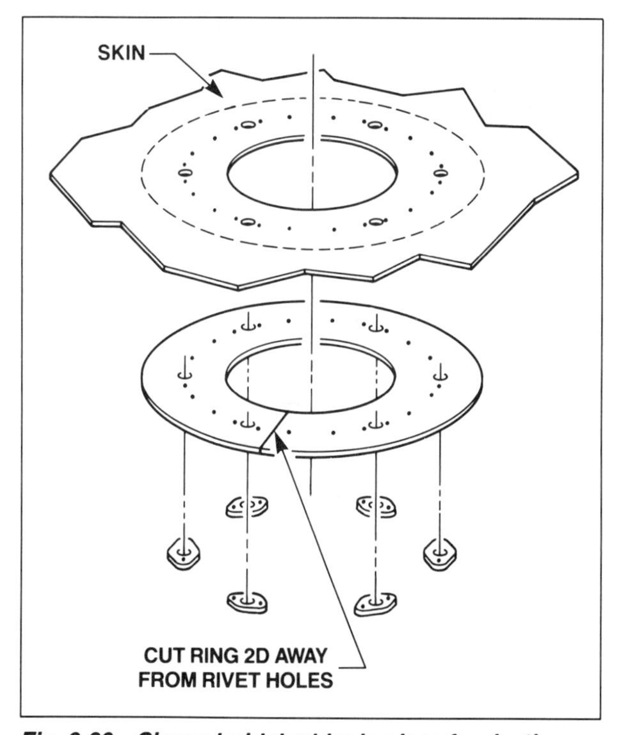

Fig. 9-26 Clecos hold doubler in place for riveting.

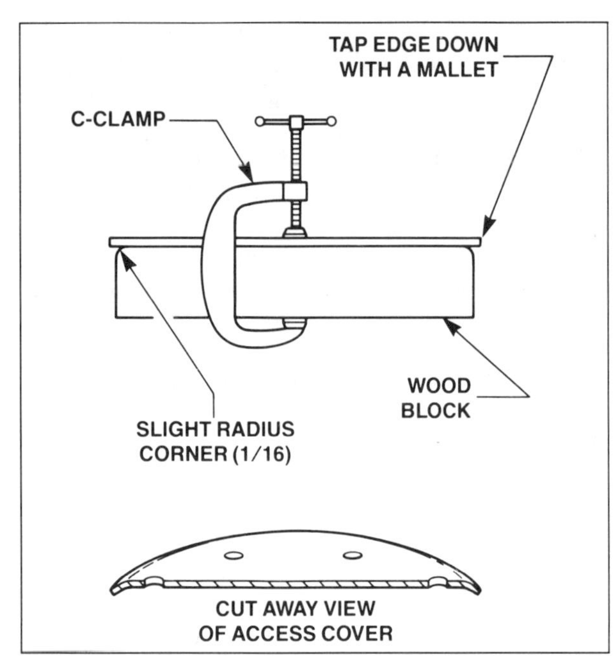

Fig. 9-27 Bent edges of inspection plate.

Please, always remember that any shortcut must still be in the best interest of aviation safety.

The value of proper rivet installation can never be taken for granted because it can and will lead to serious consequences later. Be sure to install the correct size rivets into structural locations and buck them to the minimums listed in the 43.13-1. See Chapter IV for more about riveting details.

QUESTIONS:

1. *What type of aluminum alloy is most commonly used in the construction of modern aircraft?*
2. *What is the name of the steel most commonly used for the construction of aircraft parts?*
3. *Which source is the most reliable when it comes to selecting the kind of metal to use when making aircraft parts?*
4. *What are two advantages of using a joggled joint?*
5. *What is the name of the bench tool used to make aluminum tubing or wing leading edges?*
6. *What is the main purpose of a nose rib?*
7. *Which kind of material is best suited for the construction of a nose rib block?*
8. *When forming a nose rib over a block, which two parts are formed first?*
9. *What does the layout of a control surface most nearly resemble?*
10. *Why are the trailing edges of some control surfaces joggled?*
11. *What is the maximum number of access plates allowed under the wing of an aircraft?*
12. *What purpose does the doubler serve on an access plate installation?*
13. *What is the main consideration regarding safety when installing rivets?*
14. *List three large bench tools mentioned in the chapter.*
15. *What is the bottom line on any job that you are doing in aviation?*

Appendix A

A-I Single Shear Strength Of Aluminum Alloy Rivet

ALLOYS	DRIVEN SHEAR STRENGTH IN KSI* F_S	RIVET DIAMETER						
		3/32	1/8	5/32	3/16	1/4	5/16	3/8
2117-T3	30	217	389	596	860	1560	2460	3510
2017-T3	38	275	492	755	1089	1971	3114	4446
2017-T31	34	246	441	675	974	1764	2786	3978
2024-T31	41	297	531	814	1175	2127	3360	4797
5056-H32	28	203	362	556	802	1452	2294	3276
7050-T73	43	311	557	854	1232	2230	3523	5031

*PSI values would read 30,000; 38,000; 34,000; 41,000; 28,000; and 43,000 respectively.

Single Shear Strength Of 100° Machine Countersunk Aluminum Alloy Rivet*

ALLOYS	DRIVEN SHEAR STRENGTH IN PSI	RIVET DIAMETER			
		3/32	1/8	5/32	3/16
2117-T3	30,000	186	331	518	745
2017-T3	34,000	206	368	574	828
2024-T3	41,000	241	429	670	966

*Shear strengths given are conservative estimates.

A-II Bend Allowance Chart

RADIUS / THICKNESS	1/32 .031	1/16 .063	3/32 .094	1/8 .125	5/32 .156	3/16 .188	7/32 .219	1/4 .250	9/32 .281	5/16 .313	11/32 .344	3/8 .375	7/16 .438	1/2 .500
.020	.062 .000693	.113 .001251	.161 .001792	.210 .002333	.259 .002874	.309 .003433	.358 .003974	.406 .004515	.455 .005056	.505 .005614	.554 .006155	.603 .006695	.702 .007795	.799 .008877
.025	.066 .000736	.116 .001294	.165 .001835	.214 .002376	.263 .002917	.313 .003476	.362 .004017	.410 .004558	.459 .005098	.509 .005657	.558 .006198	.607 .006739	.705 .007838	.803 .008920
.028	.068 .000759	.119 .001318	.167 .001859	.216 .002400	.265 .002941	.315 .003499	.364 .004040	.412 .004581	.461 .005122	.511 .005680	.560 .006221	.609 .006762	.708 .007853	.804 .007862
.032	.071 .000787	.121 .001345	.170 .001886	.218 .002427	.267 .002968	.317 .003526	.366 .004067	.415 .004608	.463 .005149	.514 .005708	.562 .006249	.611 .006789	.710 .007889	.807 .008971
.038	.075 .000837	.126 .001396	.174 .001937	.223 .002478	.272 .003019	.322 .003577	.371 .004118	.419 .004659	.468 .005200	.518 .005758	.567 .006299	.616 .006840	.715 .007940	.812 .009021
.040	.077 .00853	.127 .001411	.176 .001952	.224 .002493	.273 .003034	.323 .003593	.372 .004134	.421 .004675	.469 .005215	.520 .005774	.568 .006315	.617 .006856	.716 .007955	.813 .009037
.051		.134 .001413	.183 .002034	.232 .002575	.280 .003116	.331 .003675	.379 .004215	.428 .004756	.477 .005297	.527 .005855	.576 .006397	.624 .006934	.723 .008037	.821 .009119
.064		.144 .001595	.192 .002136	.241 .002676	.290 .003218	.340 .003776	.389 .004317	.437 .004858	.486 .005399	.536 .005957	.585 .006498	.634 .007039	.732 .008138	.830 .009220
.072			.198 .002202	.247 .002743	.296 .003284	.346 .003842	.385 .004283	.443 .004924	.492 .005465	.542 .006023	.591 .006564	.639 .007105	.738 .008205	.836 .009287
.078			.202 .002247	.251 .002787	.300 .003327	.350 .003885	.399 .004426	.447 .004963	.496 .005512	.546 .006070	.595 .006611	.644 .007152	.742 .008243	.840 .009333
.081			.204 .002270	.253 .002811	.302 .003351	.352 .003909	.401 .004449	.449 .004969	.498 .005535	.548 .006094	.598 .006635	.646 .007176	.744 .008266	.842 .009357
.091			.212 .002350	.260 .002891	.309 .003432	.359 .003990	.408 .004531	.456 .005072	.505 .005613	.555 .006172	.604 .006713	.653 .007254	.752 .008353	.849 .009435
.094			.214 .002374	.262 .002914	.311 .003455	.361 .004014	.410 .004555	.459 .005096	.507 .005637	.588 .006195	.606 .006736	.655 .007277	.754 .008376	.851 .009458
.102				.268 .002977	.317 .003518	.367 .004076	.416 .004617	.464 .005158	.513 .005699	.563 .006257	.612 .006798	.661 .007339	.760 .008439	.857 .009521
.109				.273 .003031	.321 .003572	.372 .004131	.420 .004672	.469 .005213	.518 .005754	.568 .006312	.617 .006853	.665 .008394	.764 .008493	.862 .009575
.125				.284 .003156	.333 .003697	.383 .004256	.432 .004797	.480 .005338	.529 .005678	.579 .006437	.628 .006978	.677 .007519	.776 .008618	.873 .009700
.156					.355 .003939	.405 .004497	.453 .005038	.502 .005579	.551 .006120	.601 .006679	.650 .007220	.698 .007761	.797 .008860	.895 .009942
.188						.417 .004747	.476 .005288	.525 .005829	.573 .006370	.624 .006928	.672 .007469	.721 .008010	.820 .009109	.917 .010191
.250								.568 .006313	.617 .06853	.667 .007412	.716 .007953	.764 .008494	.863 .009593	.961 .010675

A-III K-Factor Chart

ANG [DEG]	K-VALUE	ANG [DEG]	K-VALUE	ANG [DEG]	K-VALUE
1	0.00873	61	0.58904	121	1.7675
2	0.01745	62	0.60086	122	1.8040
3	0.02618	63	0.61280	123	1.8418
4	0.03492	64	0.62487	124	1.8807
5	0.04366	65	0.63707	125	1.9210
6	0.05241	66	0.64941	126	1.9626
7	0.06116	67	0.66188	127	2.0057
8	0.06993	68	0.67451	128	2.0503
9	0.07870	69	0.68728	129	2.0965
10	0.08749	70	0.70021	130	2.1445
11	0.09629	71	0.71329	131	2.1943
12	0.10510	72	0.72654	132	2.2460
13	0.11393	73	0.73996	133	2.2998
14	0.12278	74	0.75355	134	2.3558
15	0.13165	75	0.76733	135	2.4142
16	0.14054	76	0.78128	136	2.4751
17	0.14945	77	0.79543	137	2.5386
18	0.15838	78	0.80978	138	2.6051
19	0.16734	79	0.82434	139	2.6746
20	0.17633	80	0.83910	140	2.7475
21	0.18534	81	0.85408	141	2.8239
22	0.19438	82	0.86929	142	2.9042
23	0.20345	83	0.88472	143	2.9887
24	0.21256	84	0.90040	144	3.0777
25	0.22169	85	0.91633	145	3.1716
26	0.23087	86	0.93251	146	3.2708
27	0.24008	87	0.94890	147	3.3759
28	0.24933	88	0.96569	148	3.4874
29	0.25862	89	0.98270	149	3.6059
30	0.26795	90	1.0000	150	3.7320
31	0.27732	91	1.0176	151	3.8667
32	0.28674	92	1.0355	152	4.0108
33	0.29621	93	1.0538	153	4.1653
34	0.30573	94	1.0724	154	4.3315
35	0.31530	95	1.0913	155	4.5107
36	0.32492	96	1.1106	156	4.7046
37	0.33459	97	1.1303	157	4.9151
38	0.34433	98	1.1504	158	5.1455
39	0.35412	99	1.1708	159	5.3995
40	0.36397	100	1.1917	160	5.6713
41	0.37388	101	1.2131	161	5.9758
42	0.38386	102	1.2349	162	6.3137
43	0.39391	103	1.2572	163	6.6911
44	0.40403	104	1.2799	164	7.1154
45	0.41421	105	1.3032	165	7.5957
46	0.42447	106	1.3270	166	8.1443
47	0.43481	107	1.3514	167	8.7769
48	0.44523	108	1.3764	168	9.5144
49	0.45573	109	1.4019	169	10.385
50	0.46631	110	1.4281	170	11.430
51	0.47697	111	1.4550	171	12.706
52	0.48773	112	1.4826	172	14.301
53	0.49858	113	1.5108	173	16.350
54	0.50952	114	1.5399	174	19.081
55	0.52057	115	1.5697	175	22.904
56	0.53171	116	1.6003	176	26.636
57	0.54295	117	1.6318	177	38.188
58	0.55431	118	1.6643	178	57.290
59	0.56577	119	1.6977	179	114.590
60	0.57735	120	1.7320	180	INFINITE

A-IV Excerpt For Tangent

DEG.	SIN	TAN	COT	COS	DEG.
0.0	0.00000	0.00000	∞	1.0000	90.0
.1	.00175	.00175	573.0	1.0000	89.9
.2	.00349	.00349	286.5	1.0000	.8
.3	.00524	.00524	191.0	1.0000	.7
.4	.00698	.00698	143.24	1.0000	.6
.5	.00873	.00873	114.59	1.0000	.5
.6	.01047	.01047	95.49	0.9999	.4
.7	.01222	.01222	81.85	.9999	.3
.8	.01396	.01396	71.62	.9999	.2
.9	.01571	.01571	63.66	.9999	89.1
22.0	0.3746	0.4040	2.475	0.9272	68.0
.1	.3762	.4061	2.463	.9265	67.9
.2	.3778	.4081	2.450	.9259	.8
.3	.3795	.4101	2.438	.9252	.7
.4	.3811	.4122	2.426	.9245	.6
.5	.3827	.4142	2.414	.9239	.5
.6	.3843	.4163	2.402	.9232	.4
.7	.3859	.4183	2.391	.9225	.3
.8	.3875	.4204	2.379	.9219	.2
.9	.3891	.4224	2.367	.9212	67.1
44.0	0.6947	0.9657	1.0355	0.7193	46.0
.1	.6959	.9691	1.0319	0.7181	45.9
.2	.6972	.9725	1.0283	.7169	.8
.3	.6984	.9759	1.0247	.7157	.7
.4	.6997	.9793	1.0212	.7145	.6
.5	.7009	.9827	1.0176	.7133	.5
.6	.7022	.9861	1.0141	.7120	.4
.7	.7034	.9896	1.0105	.7108	.3
.8	.7046	.9930	1.0070	.7096	.2
.9	.7059	.9965	1.0035	.7083	45.1
45.0	0.7071	1.0000	1.0000	0.7071	45.0
DEG.	COS	COT	TAN	SIN	DEG.

A-V Rivets Per Inch Chart

NUMBER OF RIVETS REQUIRED FOR SPLICES (SINGLE-LAP JOINT) IN BARE 2014-T6, 2024-T3, 2024-T36 AND 7075-T6 SHEET, CLAD 2014-T6, 2024-T3, 2024-T36 AND 7075-T6 SHEET, 2024-T4 AND 7075-T6 PLATE, BAR, ROD, TUBE, AND EXTRUSIONS, 2014-T6 EXTRUSIONS.

THICKNESS "T" IN INCHES	NO. OF 2117-AD PROTRUDING HEAD RIVETS REQUIRED PER INCH OF WIDTH "W"					NO. OF BOLTS
	3/32	1/8	5/32	3/16	1/4	AN-3
0.016	6.5	4.9	—	—	—	—
.020	6.9	4.9	3.9	—	—	—
.025	8.6	4.9	3.9	—	—	—
.032	11.1	6.2	3.9	3.3	—	—
.036	12.5	7.0	4.5	3.3	2.4	—
.040	13.8	7.7	5.0	3.5	2.4	3.3
.051	—	9.8	6.4	4.5	2.5	3.3
.064	—	12.3	8.1	5.6	3.1	3.3
.081	—	—	10.2	7.1	3.9	3.3
.091	—	—	11.4	7.9	4.4	3.3
.102	—	—	12.8	8.9	4.9	3.4
.128	—	—	—	11.2	6.2	3.2

NOTES:

A. FOR STRINGERS IN THE UPPER SURFACE OF A WING, OR IN A FUSELAGE, 80 PERCENT OF THE NUMBER OF RIVETS SHOWN IN THE TABLE MAY BE USED.

B. FOR INTERMEDIATE FRAMES, 60 PERCENT OF THE NUMBER SHOWN MAY BE USED.

C. FOR SINGLE LAP SHEET JOINTS, 75 PERCENT OF THE NUMBER SHOWN MAY BE USED.

ENGINEERING NOTES: THE ABOVE TABLE WAS COMPUTED AS FOLLOWS:

1. THE LOAD PER INCH OF WIDTH OF MATERIAL WAS CALCULATED BY ASSUMING A STRIP ONE INCH WIDE IN TENSION.
2. NUMBER OF RIVETS REQUIRED WAS CALCULATED FOR 2117-AD RIVETS, BASED ON A RIVET ALLOWABLE SHEAR STRESS EQUAL TO 40 PERCENT OF THE SHEET ALLOWABLE TENSILE STRESS, AND A SHEET ALLOWABLE BEARING STRESS EQUAL TO 160 PERCENT OF THE SHEET ALLOWABLE TENSILE STRESS, USING NOMINAL HOLE DIAMETERS FOR RIVETS.
3. COMBINATIONS OF SHEET THICKNESS AND RIVET SIZE ABOVE THE HEAVY LINE ARE CRITICAL IN (I.E., WILL FAIL BY) BEARING ON THE SHEET; THOSE BELOW ARE CRITICAL IN SHEARING OF THE RIVETS.
4. THE NUMBER OF AN-3 BOLTS REQUIRED BELOW THE HEAVY LINE WAS CALCULATED BASED ON A SHEET ALLOWABLE TENSILE STRESS OF 70,000 PSI AND A BOLT ALLOWABLE SINGLE SHEAR LOAD OF 2,126 POUNDS.

   NOTE: Chart from AC 43.13-1A & 2A - Figure 2.28

A-VI Minimum Bend Radius Chart

RECOMMENDED RADII FOR 90° BENDS IN ALUMINUM ALLOYS						
ALLOY AND TEMPER	APPROXIMATE SHEET THICKNESS (t) (INCH)					
	0.016	0.032	0.064	0.128	0.182	0.258
2024-0[1]	0	0—1t	0—1t	0—1t	0—1t	0—1t
2024-T3[1,2]	1-1/2t—3t	2t—4t	3t—5t	4t—6t	4t—6t	5t—7t
2024-T6[1]	2t—4t	3t—5t	3t—5t	4t—6t	5t—7t	6t—10t
5052-0	0	0	0—1t	0—1t	0—1t	0—1t
5052-H32	0	0	1/2t—1t	1/2t—1-1/2t	1/2t—1-1/2t	1/2t—1-1/2t
5052-H34	0	0	1/2t—1-1/2t	1-1/2—2-1/2t	1-1/2—2-1/2t	2t—3t
5052-H36	0—1t	1/2t—1-1/2t	1t—2t	1-1/2t—3t	2t—4t	2t—4t
5052-H38	1/2t—1-1/2t	1t—2t	1-1/2t—3t	2t—4t	3t—5t	4t—6t
6061-0	0	0—1t	0—1t	0—1t	0—1t	0—1t
6061-T4	0—1t	0—1t	1/2t—1-1/2t	1t—2t	1-1/2t—3t	2-1/2t—4t
6061-T6	0—1t	1/2t—1-1/2t	1t—2t	1-1/2t—3t	2t—4t	3t—4t
7075-0	0	0—1t	0—1t	1/2t—1-1/2t	1t—2t	1-1/2t—3t
7075-T6[1]	2t—4t	3t—5t	4t—6t	5t—7t	5t—7t	6t—10t

[1]ALCLAD SHEET MAY BE BENT OVER SLIGHTLY SMALLER RADII THAN THE CORRESPONDING TEMPERS OF UNCOATED ALLOY.

[2]IMMEDIATELY AFTER QUENCHING, THIS ALLOY MAY BE FORMED OVER APPRECIABLY SMALLER RADII.

Appendix B

B-I Introduction To A Computer Program For Computing Bend Allowance, Setback And Layout

The rising cost of a new airplane is the main reason why many owners refurbish, rebuild and recondition their old aircraft. New aircraft parts are available for late models, but for many older aircraft, parts have to be made in sheet metal or machine shops.

The computer program in Appendix B-II enables the technician to compute bend allowance, setback, and layout of any part with a bend angle of 1 to 180 degrees. The program is written in BASIC language so that it can be used in a variety of models of personal computers (PCs). To be able to use this program, the operator must know the thickness of the metal, the bend radius and the number of degrees of bend. In addition, the mechanic must know the various leg lengths. The program can be used to make "L" or "U" shaped channel layouts. This means stringers, longerons or spars can be computed using the data entered into this program.

The program will produce all the mathematical solutions needed to make a layout. The only thing the mechanic will have to do is place the sight line in the correct location of the bend allowance area. Following the listing of the program are several examples of computer layouts.

It should be noted that when you copy this program into your PC computer, the print statements must be enclosed by double quotes (") or whatever your BASIC computer language requires. After copying this program into your computer, be sure to type in the command SAVE. If you want to execute a problem, type in or use a function key to indicate RUN.

The program begins with a series of information instructions. Next, it will stop and wait for you to enter the metal thickness. Simply type in whatever thickness (in thousandths of an inch) you are planning to use. The computer will then move on to the next question and ask for the bend radius (also done in thousandths of an inch). The next step will ask for the bend angle. After this question, the bend allowance portion of the program will be loaded and the computer will begin to ask questions related to computing setback.

To calculate setback for bends less than 90 degrees, the operator will need a K-Factor chart (Appendix B-III). The K-Factor is a function of tangent taken from a trigonometric function table. For example, if the bend is 45 degrees, divide it by two; it equals 22.5 degrees. If you consult a trigonometric function table, you will find 22.5 degrees is equal to .4142.

For bends 90 degrees or greater, Appendix B-III shows that setback uses a K-Factor of one. If you wanted to do a mathematical setback, the K-Factor would be greater than one.

When laying out an aircraft part in the shop, we are not interested in what the distance is from the outside mold point. Therefore, setback is found by subtracting radius and thickness from the finished leg length to find the unbent portion. The computer program takes these two options into account while executing setback.

After the computer has calculated bend allowance from the input date of radius, thickness and number of degree bend, another series of questions is asked of the operator. The next question asked is the K-Factor for the left side bend. You can enter a K-Factor of "1" for bends 90 degrees and greater, or the appropriate selection from a K-Factor chart for bends less than 90 degrees. A similar response is requested if there is another leg required. After the setback data is entered, the computer then moves into the final phase of computing layout.

The next three questions asked by the computer relate to the individual leg lengths of the finished part. Notice in Appendix B-V that the legs are called A, B and C. A is the left side leg, B is the horizontal leg, and C is the right side leg. The computer will ask for the leg length of A. You may type in whatever finished length is required to make this part. Next it will ask for the leg length of B, then the leg length of C.

B-I (cont'd.)

Prior to typing in the length of C, the computer will instruct you to turn on your printer. Turning on your printer will give you a printout of each leg and bend allowance so you can make a layout. The program concludes by asking if you want to do another problem. A "Y" response will take you back to the start of the program for another RUN.

The only step which the computer cannot do is to locate the sight line on your layout. You can do that by placing the sight line one bend radius from the bend line positioned under the nose of the brake.

B-II A Computer Program For Computing Bend Allowance, Setback And Layout

```
1 PRINT "A COMPUTER PROGRAM FOR CALCULATING BEND ALLOWANCE, SETBACK,"
2 PRINT "AND LAYOUT FOR BENDS FROM 1 TO 180 DEGREES."
3 PRINT
4 PRINT
5 PRINT TAB(25), "BY-NICK BONACCI PROFESSOR OF AVIATION LEWIS UNIVERSITY"
6 PRINT
7 PRINT "ENTER BEND RADIUS THAT YOU WILL BE USING -->";
8 INPUT R
9 PRINT
10 PRINT "ENTER THICKNESS OF THE METAL THAT YOU ARE USING -->";
11 INPUT T
12 PRINT
13 PRINT "ENTER THE NO. OF DEGREES USED ON LEFT SIDE -->";
14 INPUT D1
15 PRINT
16 PRINT "ENTER THE NO OF DEGREES USED ON RIGHT SIDE -->";
17 INPUT D2
18 PRINT
19 X=(.01743*R+.0078*T)
20 PRINT "BEND ALLOWANCE FOR ONE DEGREE OF BEND =";X
21 B=X*D1
22 PRINT
23 PRINT "BEND ALLOWANCE =";B
24 B1=X*D2
25 PRINT
26 PRINT "BEND ALLOWANCE =";B1
27 PRINT
28 PRINT "WHAT IS THE K-FACTOR OF LEFT SIDE ANGLE -->";
29 INPUT K
30 S=K*(R+T)
31 PRINT
32 PRINT "SET BACK FOR LEFT SIDE ANGLE =";S
33 PRINT
34 PRINT "WHAT IS THE K-FACTOR OF RIGHT SIDE ANGLE -->";
35 INPUT K1
36 PRINT
37 S1=K1*(R+T)
38 PRINT "SET BACK FOR RIGHT SIDE ANGLE =";S1
39 PRINT
40 PRINT "ENTER FINISHED LENGTH OF LEG (A) -->";
41 INPUT W
42 PRINT
43 PRINT "ENTER FINISHED LENGTH OF LEG (B) -->";
44 INPUT Y
45 PRINT
46 PRINT " TURN ON PRINTER-ENTER FINISHED LENGTH OF LEG (C) --.";
47 INPUT Z
48 PRINT
49 W1=W-S
50 PRINT "---------------------------------------"
51 PRINT "UNBENT PORTION FOR LEFT SIDE LEG A =";W1
52 PRINT
53 PRINT "LEFT SIDE BEND ALLOWANCE IS-->";B
54 PRINT
55 Y1=Y-S-S1
56 PRINT "UNBENT PORTION FOR FLAT LEG B =";Y1
57 Z1=Z-S1
58 PRINT
59 PRINT "RIGHT SIDE BEND ALLOWANCE IS -->";B1
60 PRINT
61 PRINT "UNBENT PORTION FOR RIGHT SIDE LEG C =";Z1
62 PRINT
63 T=W1+B+Y1+B1+Z1
64 PRINT
65 PRINT "TOTAL LENGTH OF LAYOUT A+B+C=";T
66 PRINT "---------------------------------------"
67 PRINT
68 PRINT "DO YOU WANT TO DO ANOTHER BEND ALLOWANCE PROBLEM"
69 PRINT
70 PRINT "WHERE YOU CAN DO ANY KIND OF BEND OR LAYOUT WITH"
71 PRINT
72 PRINT "THREE LEGS OR LESS (Y/N)";
73 INPUT A$
74 IF A$="Y" GOTO 1
75 IF A$="N" GOTO 77
76 PRINT
77 END
```

Example Of Computer Program Runout

```
ENTER BEND RADIUS THAT YOU WILL BE USING -->? .250

ENTER THICKNESS OF THE METAL THAT YOU ARE USING -->? .051

ENTER THE NO. OF DEGREES USED ON LEFT SIDE -->? 90

ENTER THE NO OF DEGREES USED ON RIGHT SIDE -->? 90

BEND ALLOWANCE FOR ONE DEGREE OF BEND = .0047553

BEND ALLOWANCE = .427977

BEND ALLOWANCE = .427977

WHAT IS THE K-FACTOR OF LEFT SIDE ANGLE -->? 1

SET BACK FOR LEFT SIDE ANGLE = .301

WHAT IS THE K-FACTOR OF RIGHT SIDE ANGLE -->? 1

SET BACK FOR RIGHT SIDE ANGLE = .301

ENTER FINISHED LENGTH OF LEG (A) -->? 2

ENTER FINISHED LENGTH OF LEG (B) -->? 2

 TURN ON PRINTER-ENTER FINISHED LENGTH OF LEG (C) --.? 2

---------------------------------------
UNBENT PORTION FOR LEFT SIDE LEG A = 1.699

LEFT SIDE BEND ALLOWANCE IS--> .427977

UNBENT PORTION FOR FLAT LEG B = 1.398

RIGHT SIDE BEND ALLOWANCE IS --> .427977

UNBENT PORTION FOR RIGHT SIDE LEG C = 1.699

TOTAL LENGTH OF LAYOUT A+B+C= 5.65194
---------------------------------------

DO YOU WANT TO DO ANOTHER BEND ALLOWANCE PROBLEM

WHERE YOU CAN DO ANY KIND OF BEND OR LAYOUT WITH

THREE LEGS OR LESS (Y/N)? Y
```

B-III Setback (K) Chart And Natural Trigonometric Functions Chart

FOR BEND OTHER THAN 90° SETBACK = [R + T]							
ANG [DEG]	K-VALUE	ANG [DEG]	K-VALUE	ANG [DEG]	K-VALUE	ANG [DEG]	K-VALUE
1	0.00873	46	0.42447	91		136	
2	0.01745	47	0.43481	92		137	
3	0.02618	48	0.44523	93		138	
4	0.03492	49	0.45573	94		139	
5	0.04366	50	0.46631	95		140	
6	0.05241	51	0.47697	96		141	
7	0.06116	52	0.48773	97		142	
8	0.06993	53	0.49858	98		143	
9	0.07870	54	0.50952	99		144	
10	0.08749	55	0.52057	100		145	
11	0.09629	56	0.53171	101		146	
12	0.10510	57	0.54295	102		147	
13	0.11393	58	0.55431	103		148	
14	0.12278	59	0.56577	104		149	
15	0.13165	60	0.57735	105		150	
16	0.14054	61	0.58904	106	K	151	K
17	0.14945	62	0.60086	107		152	
18	0.15838	63	0.61280	108	F	153	F
19	0.16734	64	0.62487	109	A	154	A
20	0.17633	65	0.63707	110	C	155	C
21	0.18534	66	0.64941	111	T	156	T
22	0.19438	67	0.66188	112	O	157	O
23	0.20345	68	0.67451	113	R	158	R
24	0.21256	69	0.68728	114		159	
25	0.22169	70	0.70021	115	O	160	O
26	0.23087	71	0.71329	116	F	161	F
27	0.24008	72	0.72654	117		162	
28	0.24933	73	0.73996	118	ONE	163	ONE
29	0.25862	74	0.75355	119		164	
30	0.26795	75	0.76733	120	1	165	1
31	0.27732	76	0.78128	121		166	
32	0.28674	77	0.79543	122		167	
33	0.29621	78	0.80978	123		168	
34	0.30573	79	0.82434	124		169	
35	0.31530	80	0.83910	125		170	
36	0.32492	81	0.85408	126		171	
37	0.33459	82	0.86929	127		172	
38	0.34433	83	0.88472	128		173	
39	0.35412	84	0.90040	129		174	
40	0.36397	85	0.91633	130		175	
41	0.37388	86	0.93251	131		176	
42	0.38386	87	0.94890	132		177	
43	0.39391	88	0.96569	133		178	
44	0.40403	89	0.98270	134		179	
45	0.41421	90	1.0000	135		180	

DEG	SIN	TAN	COT	COS	DEG
30.0	0.5000	0.5774	1.7321	0.8660	60.0
.1	.5015	.5797	1.7251	.8652	59.9
.2	.5030	.5820	1.7182	.8643	.8
.3	.5045	.5844	1.7113	.8634	.7
.4	.5060	.5867	1.7045	.8625	.6
.5	.5075	.5890	1.6977	.8616	.5
.6	.5090	.5914	1.6909	.8607	.4
.7	.5105	.5938	1.6842	.8599	.3
.8	.5120	.5961	1.6775	.8590	.2
.9	.5135	.5985	1.6709	.8581	59.1

60 ÷ 2 = 30
K = .5774

DEG	SIN	TAN	COT	COS	DEG
20.0	0.3420	0.3640	2.747	0.9397	70.0
.1	.3437	.3659	2.733	.9391	69.9
.2	.3453	.3679	2.718	.9385	.8
.3	.3469	.3699	2.703	.9379	.7
.4	.3486	.3719	2.689	.9373	.6
.5	.3502	.3739	2.675	.9367	.5
.6	.3518	.3759	2.660	.9361	.4
.7	.3535	.3779	2.646	.9354	.3
.8	.3551	.3799	2.633	.9348	.2
.9	.3567	.3819	2.619	.9342	69.1

40 ÷ 2 = 20
K = .3640

DEG	SIN	TAN	COT	COS	DEG
22.0	0.3746	0.4040	2.475	0.9272	68.0
.1	.3762	.4061	2.463	.9265	67.9
.2	.3778	.4081	2.450	.9259	.8
.3	.3795	.4101	2.438	.9252	.7
.4	.3811	.4122	2.426	.9245	.6
.5	.3827	.4142	2.414	.9239	.5
.6	.3843	.4163	2.402	.9232	.4
.7	.3859	.4183	2.391	.9225	.3
.8	.3875	.4204	2.379	.9219	.2
.9	.3891	.4224	2.367	.9212	67.1

45 ÷ 2 = 22.5
K = .4142

B-IV Computation Of Setback

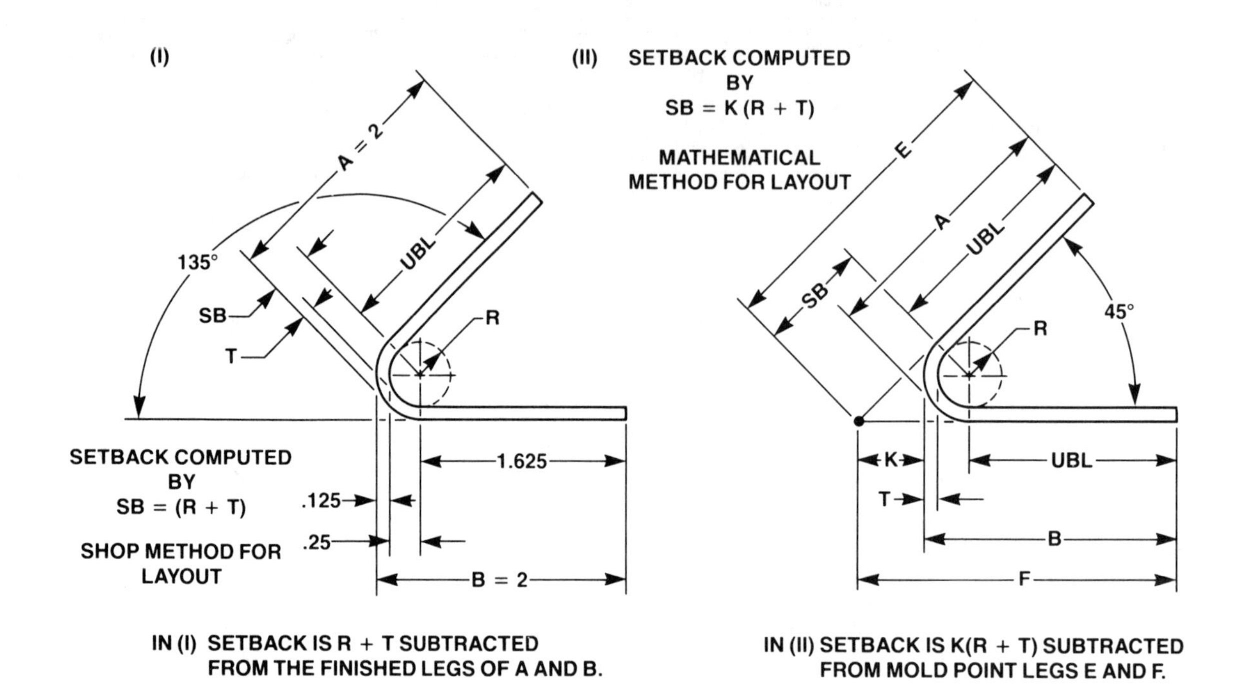

IN (I) SETBACK IS R + T SUBTRACTED FROM THE FINISHED LEGS OF A AND B.

IN (II) SETBACK IS K(R + T) SUBTRACTED FROM MOLD POINT LEGS E AND F.

WHEN COMPUTING LAYOUT LEGS E AND F ARE NOT PRACTICAL TO USE. THEREFORE THIS METHOD IS NOT ACCEPTABLE.

B-V Labeling Of Leg Lengths For Computer Setback

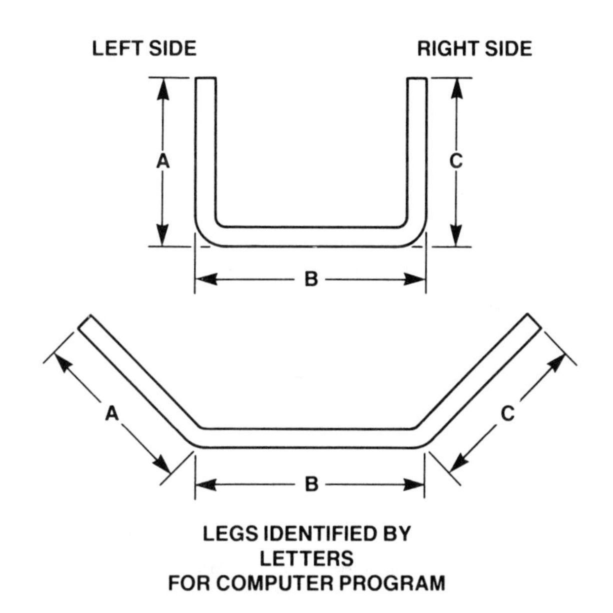

LEGS IDENTIFIED BY
LETTERS
FOR COMPUTER PROGRAM

B-VI Computing Shear Strength Using An IBM XT In Basic Language

```
5 PRINT          "DRIVEN SHEAR STRENGTH OF A RIVET"
10 PRINT
15 PRINT
17 PRINT "COPY THE DRILL SIZE FROM #40=.098,#30=.1285,#21=.159,#11=.191"
18 PRINT
20 PRINT "ENTER DRILL DIAMETER IN THOUSANDS OF AN INCH -->";
25 INPUT D
30 PRINT
35 LET A=3.1416*((.5*D)^2)
50 PRINT
55 PRINT "USE ULTIMATE SHEAR STRENGTH TAKEN FROM THE LIST BELOW -->"
56 PRINT
57 PRINT "ULTIMATE SHEAR OF 2117T3=30KSI,2017T3=34KSI&2024T31=41KSI-->"
58 PRINT "NOTE: CHANGE KSI TO PSI BY MULTIPLYING KSI X 1000-->";
59 PRINT
60 PRINT "ENTER ULTIMATE SHEAR STRENGTH"
65 INPUT U
67 PRINT
70 PRINT "ENTER NUMBER OF SHEARING PLANES-->";
75 INPUT P
80 LET S=A*U*P
90 PRINT S
93 PRINT
94 PRINT "____________________________________________________________"
95 PRINT "THE SHEAR STRENGTH OF THE RIVET YOU SELECTED IS-->";S
96 PRINT "____________________________________________________________"
97 PRINT
100 PRINT "DO YOU WANT TO DO ANOTHER RIVET SHEAR PROBLEM <Y/N>";
105 INPUT X$
110 IF X$="Y" GOTO 17
115 IF X$="N" GOTO 130
120 GOTO 100
130 END
```

Example Of Computer Program Runout

```
RUN
DRIVEN SHEAR STRENGTH OF A RIVET


COPY THE DRILL SIZE FROM #40=.098,#30=.1285,#21=.159,#11=.191

ENTER DRILL DIAMETER IN THOUSANDS OF AN INCH -->? .1285


USE ULTIMATE SHEAR STRENGTH TAKEN FROM THE LIST BELOW -->

ULTIMATE SHEAR OF 2117T3=30KSI,2017T3=34KSI&2024T31=41KSI-->
NOTE: CHANGE KSI TO PSI BY MULTIPLYING KSI X 1000-->
ENTER ULTIMATE SHEAR STRENGTH
? 30000

ENTER NUMBER OF SHEARING PLANES-->? 1
 389.0616

____________________________________________________________
THE SHEAR STRENGTH OF THE RIVET YOU SELECTED IS--> 389.0616
____________________________________________________________

DO YOU WANT TO DO ANOTHER RIVET SHEAR PROBLEM <Y/N>?
```

B-VII Selecting Correct Rivet Diameter Using An IBM Computer

```
5 PRINT TAB(25), "DIAMETER OF A RIVET USED FOR A REPAIR"
10 PRINT
15 PRINT
20 PRINT "ENTER THE THICKNESS OF THE METAL TO BE RIVETED-->";
25 INPUT T
30 D=T*3
35 PRINT D
40 IF D>=.188 GOTO 95
45 IF D>=.153 GOTO 105
50 IF D>=.135 GOTO 105
55 IF D>=.12 GOTO 115
60 IF D>=.108 GOTO 115
65 IF D>=.096 GOTO 115
70 IF D>=.075 GOTO 125
75 IF D<=.096 GOTO 125
80 IF D=.045 GOTO 125
90 PRINT
95 PRINT "USE A 3/16 RIVET"
96 PRINT
100 GOTO 140
105 PRINT "USE A 5/32 RIVET"
106 PRINT
110 GOTO 140
115 PRINT "USE A 1/8 RIVET"
116 PRINT
120 GOTO 140
125 PRINT "USE A 3/32 RIVET"
126 PRINT
130 GOTO 140
135 PRINT
140 PRINT "DO YOU WANT TO DO ANOTHER PROBLEM (Y/N)";
141 PRINT
142 PRINT
145 INPUT X$
150 IF X$= "Y" GOTO 15
151 IF X$= "y" GOTO 15
155 IF X$= "N" GOTO 165
160 PRINT
165 END
```

Example Of Computer Program Runout

```
Version A3.10 Copyright IBM Corp. 1981, 1985
60882 Bytes free

Ok
LOAD"r diam.asc
Ok
RUN
                                        DIAMETER OF A RIVET USED FOR A REPAIR


ENTER THE THICKNESS OF THE METAL TO BE RIVETED-->? .040
 .12
USE A 1/8 RIVET
DO YOU WANT TO DO ANOTHER PROBLEM (Y/N)? y

Ok
RUN
                                        DIAMETER OF A RIVET USED FOR A REPAIR

ENTER THE THICKNESS OF THE METAL TO BE RIVETED-->? .051
 .153
USE A 5/32 RIVET
DO YOU WANT TO DO ANOTHER PROBLEM (Y/N)?
1LIST   2RUN    3  L  O  A  D  "                      4  S  A  V  E  "
O  N  8  T  R  O  F  F  K  E     Y
```

Glossary Of Terms

AC 43.13-1A An advisory circular in book form, issued by the Federal Aviation Administration, which covers acceptable methods, techniques, and practices for aircraft inspection and repair.

Airloc A fastener used to lock engine cowling or inspection plates in place by use of a spring and lock pin.

Alclad A coating of pure aluminum applied to the surface of an alloy to make it corrosion resistant.

Alodining A treatment applied to the surface of aluminum alloy to prevent corrosion and provide an excellent paint base.

Alodizing A chemical surface treatment to prepare aluminum for finish painting.

Aluminum alloy A physical mixture of metals in which aluminum is the base and other metals are added for specific characteristics.

Annealing Heating of an alloy to a temperature called solid solution temperature, allowing it to cool slowly at a controlled rate through its critical range for the purpose of inducing softness. Results in removal of former heat-treatment strain hardening and internal stress.

Anode The positive pole of an electrical device. In the anodizing process, the aluminum part is the anode.

Anodizing An electrochemical action which causes aluminum oxides to solidify on the surfaces of aluminum alloy.

Anti-drag wire A wire brace acting in a diagonal direction that opposes a drag wire.

Ballizing The forcing of a ball into an undersize hole, causing deformation and plastic recovery to occur. The result is a hardened cylindrical wall that has high resistance to bearing and shearing failure. The area around the hole becomes preloaded.

Bearing load The load applied by one item placed on, or clamped between, another item; e.g. the load applied to the structure under a rivet head or bolt head.

Beef-up To strengthen an area which is showing signs of fatigue failure.

Beehive spring The coiled set retainer spring of a pneumatic rivet gun.

Bend allowance The amount of material actually used in a bend of sheet metal.

Bending The stresses in an object caused by a load being applied to the one end while the other is restrained. This results in a tensile load on one side and a compressive load on the other.

Bend radius The radius of the inside of a bending sheet metal.

Bend tangent A line made on sheet metal layout to indicate the beginning of a bend.

Bi-metallic The fusion of two different metals on the same primary member; e.g. a Cherrybuck rivet.

Boeing rivet A modified AN470 rivet.

Brazier head rivet An aircraft rivet with a large, thin head, specification AN455. This rivet has been superseded by the AN470 universal head rivet.

Bucking bar A hard, steel bar having a polished face, used to form the upset head of an aircraft.

Bucktail The driven end of a solid-shank rivet. Also called shop head, upset head, driven head, or bucked head.

Bulbed Cherrylock rivet A special blind rivet manufactured by Cherry, in which the stem is locked into the hollow shank by a locking collar which swages into a groove in the stem.

Bulbed end A formed end on a blind rivet which is inaccessible during installation.

Caliper, inside A measuring instrument with two adjustable legs, used to determine an inside measurement. Once the distance has been established, the actual measurement is made with a steel scale or a micrometer or vernier caliper.

Caliper, micrometer A precision measuring device having a single movable jaw, advanced by a screw. One revolution of the screw advances the jaw 0.025 inch.

Caliper, outside A measuring device having two movable legs, used to determine the distance across an object. Once the distance has been established, the actual dimension may be made by using a steel scale or a vernier caliper.

Caliper, vernier A precision measuring tool used to measure the inside or outside dimension of an object. An auxiliary, or vernier, scale is used to accurately divide the increments of the regular scale.

Caliper, vernier micrometer A micrometer caliper with a special vernier scale which allows each one-thousandth-inch increment to be broken down into ten equal parts so that one ten-thousandth of an inch may be accurately read.

Camloc A fastener used to hold cowling in place by a lock-pin and spring assembly.

Cantilever A beam or other member supported at, or near, one end only, without external bracing.

Cap strip A reinforcing, or forming, piece of material used principally on wing beams (spars) or wing ribs.

Cathode The negative pole of an electrical device. In the anodizing process, the steel holding tank is the cathode.

Channel A metal structural member either extruded or bent into a U-shape.

Chromally A nickname for the steel alloy chrome-molybdenum.

Clad aluminum Aluminum alloy which has a coating of pure aluminum rolled onto both sides for corrosion protection.

Clamp-up The drawing together of two or more sheets of aluminum skin and parts.

Cleco A sheet metal clamp used to hold parts in place prior to driving the rivets.

Coldworking Any mechanical process which increases metal hardness. May be accomplished by hammering, rolling material through rollers, or pulling through dies.

Compound curve Curvature of a metal surface in more than one plane.

Compression The resultant of two forces which act along the same line, and act towards each other.

Compression member A heavy member, usually of tubular steel, which separates the spars in a Pratt truss wing, and is used to carry only compression loads.

Compression rib A heavy duty rib specially made with heavy cap strips and extra strength webs; designed to withstand compression loads between wing spars.

Corrosion An electrolytic action in a metal which reduces the metal to porous or granular salts.

Corrugated skin A thin sheet metal skin for aircraft structure. V's or a regular wavy surface increase the rigidity of the metal.

Countersinking Making a conical hole in the surface of an aircraft skin so that it can receive a flush rivet or screw. The conical hole is called a nest or well.

CRES Corrosion resistant steel or stainless steel. The letters are assigned to parts and fasteners.

Crocus cloth Abrasive fabric having a fine surface coating of red ferric oxide, used for polishing metals.

Deburring Removal of slivers or rough edges from the area around a drilled hole prior to installing rivets or fasteners.

De-pinning tool A pick used to remove material lodged in the gullets of a file.

Dimpling A process whereby thin sheet metal is recessed into a dimpling die to form a depression for a flush rivet.

Dimpling coin A form of dimpling in which the metal is normally heated and forced around the male die by a ramming action of the female die. This produces very sharp corners in the dimple.

Dimpling radius A form of cold-dimpling of thin sheet metal in which the cone-shaped male die is forced into the recess of the female die with either a hammer blow or a pneumatic rivet gun.

Doubler A reinforcement which adds stiffening to the area surrounding a repair. Thin wing skins require a doubler when patches are installed.

Drag wire A diagonal, load-carrying member of a Pratt truss wing. It runs from the front spar inboard to the rear spar outboard, and carries tensile loads that tend to drag backward on the wing.

Ductility The ability of a metal to be drawn into shapes, such as wires, bars, or extrusions.

Duralumin the original name of the aluminum alloy now known as 2017. It was first produced in Germany and used in their great Zeppelin fleet of World War I.

Dzus fastener A fastener used to hold an engine cowling in place by locking a stud into a lockspring assembly.

Edge distance The distance between the center of the rivet hole and the edge of the material.

Empennage The tail section of an aircraft, which includes the horizontal stabilizer, elevator, rudder, and vertical stabilizer.

Exfoliation A form of intergranular corrosion that attacks extruded metals along their layer-like grain structure.

Explosive rivet A patented blind rivet manufactured by the DuPont Company. Its hollow end is filled with an explosive and sealed with a plastic cap. When the rivet is heated, it explodes, swelling its end and clamping the metal together.

Extrusion A strip of metal, usually of aluminum or magnesium, which has been forced in its plastic state through a die. This can produce complex cross-sectional shapes required for modern aircraft production.

FAA Federal Aviation Administration.

FAA-approved data Data which may be issued as authorization for the techniques or procedures for making a repair or an alteration to certificated aircraft. Approved data may consist of such documents as Advisory Circular 43.13-1A and -2A; Manufacturer's Service Bulletins; manufacturer's kit instructions; Airworthiness Directives; or specific details of a repair issued by the engineering department of the manufacturer.

FAA Form 337 Major Repair and Alteration form.

Fairing A metal part used to streamline the wings to the fuselage for smoother air flow during flight.

Fayed edges Two sheets of metal butted together and riveted into a spar, stringer, or rib.

Firewall On light aircraft, a wall that is located between the engine compartment and the cockpit. On multi-engine aircraft, the firewall is located between the engine compartment and the wheel well.

Flush rivet A countersunk rivet in which the manufactured head is flush with the surface of the metal when it is properly driven.

Fly cutter A cutting tool used to cut round holes in sheet metal. It is turned by a drill press, and the cutting is done by a tool bit held in an adjustable arm.

Formability The ability of metal to be formed or bent to minimum radius without cracking.

Formers Frames of light wood or metal which attach to the truss of the fuselage or wing in order to provide the required aerodynamic shape.

Forming block A block, usually made of hardwood, around which metal parts are formed.

Galvanic action Electrical pressure within a substance which causes electron flow due to differences of electrode potential within the material.

Grip length The length of the unthreaded shank of a blind rivet between the manufactured head and the maximum extent of the pulled head. The maximum thickness of material that can be joined by a fastener.

Hi-shear rivet A special form of threadless bolt used for high-speed, high-strength, lightweight construction of an aircraft. A steel pin is held into the structure by an aluminum or mild-steel collar, swaged into a groove around the end of the pin.

High-speed drill A drill which can withstand high temperature caused by friction from high drill tip speed.

Hook rule A steel scale with a hook, or projection on one end so the rule can be used to measure to the edge of materials having edge radii.

Huck lockbolt A patented threadless bolt used in the production of aircraft where high strength, high speed, lightweight fasteners are required.

Icebox rivet A rivet made of 2024 or 2017 aluminum alloy, which must be heat treated, quenched, and held in a sub-zero ice box until it is driven.

Included angle The total number of degrees in an angle, as measured from 0° to 360°; e.g. a 270° included angle.

Jig A framework or alignment structure used in the construction of an aircraft to hold all the parts in proper alignment while they are fastened together.

Joggle A small offset in sheet metal formed to allow one part to overlap another.

K factor A function of tangent. It is used to compute the radius and thickness of bend with angles less than 90 degrees.

Krueger flap A leading edge wing flap hinged at the bottom side of the airfoil. When actuated, the leading edge bends downward, increasing the camber.

KSI Kilo square inch. One KSI = 1,000 PSI.

Lap joint A sheet metal structural joint created by overlapping sheet edges.

Leaf brake A bending tool used to form straight bends in sheet metal. The material is clamped in the brake, and a heavy leaf folds the metal back over a radius block to form the desired bend.

Locked spindle A device which locks the center stem inside a blind rivet shank.

Longeron The primary structural member of a fuselage. It runs fore and aft.

Malleable A type of material, normally a metal, which can be shaped by hammering.

Mandrel An axle, spindle, or arbor inserted into a hole in a piece of work to support it during machining. Or, a metal bar that serves as a core around which material can be cast, molded, forged, bent or shaped. Or, the shaft and bearings on which a tool is mounted.

Mold line A line formed by the intersection of the flat surfaces of two sides of a sheet metal part.

Monocoque A stressed-skin type of construction in which the stiffness and shape of the skin provide a large measure of the strength of the structure. No truss or sub-structure is required.

NACA National Advisory Commission of Aeronautics.

Nutplate A special form of nut which may be riveted to the inside of a structure.

Oil canning A condition of sheet metal skin which is slightly bulged or stretched between rows of rivets. The bulge will pop back and forth in the manner of an oil can.

Open angle A bend of less than 90 degrees.

Pratt truss A type of truss structure in which the vertical members carry only compressive loads, and the diagonal members carry only tensile loads. A Pratt truss is used for most fabric covered wings.

Precipitation heat treatment A step in the heat treating process of aluminum in which the metal, after having been heated to its critical temperature and quenched, is raised to an elevated temperature and held for a period of time. This process artificially ages the metal and increases its strength.

Protruding head rivet An aircraft rivet in which the head protrudes above the surface of the metal. Universal-head, round-head, and flat-head rivets are all forms of protruding head rivets.

Pull-together The action of a blind fastener when it draws two or more sheets of metal together.

Radius block A metal block around which sheet metal is bent to obtain a specific bend radius.

Radius gage A precision gage having accurately cut inside and outside radii, used to measure the radius of a bend.

Rawhide mallet A mallet made of rawhide, wound into a tight cylinder, used to form sheet metal without denting it.

Reactive metal A metal, such as aluminum or magnesium, which readily reacts with oxygen to form corrosion.

Rivet gage The transverse pitch, or distance between rows of rivets.

Rivet pitch The distance between rivets in a row.

Rivet set The tool which fits into a rivet gun, used to hammer against the manufactured head of a rivet so the bucking bar may form the upset head on the opposite side of the skin.

Rivet snap Another name for rivet set.

Rivnut A patented, hollow, blind rivet, manufactured by the BF Goodrich Company, in which the inside of the shank is threaded. The upset rivet may be used as a blind nut.

Screw-jack assembly A mechanical device for transmitting linear motion, as in a flap drive or stabilizer trim assembly.

Scriber A hardened-steel or carbide-tipped sharp-pointed tool, used to scribe lines on metal for cutting.

Setback The distance between the mold line and the bend tangent line on a sheet metal layout.

Semicantilever A wing which uses external bracing.

Semimonocoque A fuselage which uses internal bracing such as bulkheads, longerons, former rings, and stringers.

Sharpening angle The established angles and relief angles on the drill; e.g. 59° included angle standard, 118° included angle for sheet metal.

Shear load The pressure required to shear or cut material, as in rivet shear.

Shear stress A stress exerted on a material which tends to slide it apart.

Slip rollers Metalworking devices used to shape sheet metal into cylindrical and curved shapes.

Solution heat treatment A form of heat treatment of aluminum alloy in which the metal is raised to its heat-treating temperature, held at this temperature until it is uniform throughout, and then quenched. This process causes the alloying agents to be held in a solid solution, which increases the strength of the metal.

Spar The principle, spanwise member of a truss-type wing structure.

Spoilers Devices on the wings of jet aircraft which assist the aileron in flight and open to prevent rebounding during landing.

Springback The number of degrees that metal will unbend after being bent.

Spotface To level a slanted surface in order to install bolts or special fasteners.

Squeeze gun a pneumatic or hydraulic riveting gun, in which a set for the manufactured head and a smooth surface to form the upset head are mounted in the jaw of the large clamp. When the squeeze gun is actuated, the jaws come together just enough to form the proper size upset head.

Stop countersink A special countersink having a collar which will not allow the cutter to cut too deep into the metal skin.

Stop-drill A hole drilled in the end of a crack in aircraft structural material to distribute the stresses and stop the crack from proceeding.

Straight-pin fastener A fastener whose shank sides are parallel.

Strain hardening The increase in strength and hardness of a metal by workhardening or cold-working. Strain hardening is normally done after a piece of material has been heat treated, and is the only way by which a non-heat-treatable aluminum alloy may be hardened.

Stress The internal force in a body that resists the tendency of an external force to change its shape.

Stringers Thin metal or wood strips running the length of the fuselage to fill in the shape of the formers.

Superstructure Areas where major stress is concentrated, such as keel beams, wing center sections, and wing spars.

Sympathetic vibration A vibration caused by an adjacent but unconnected item.

Tensile strength Strength of an item/material measured by pulling apart a sample of the item/material.

Tension Stress produced in a body by forces acting along the same line but in opposite directions.

Thin-sheet take-up The drawing together of skins which occurs when special fasteners are installed.

Throatless shears A heavy-duty shear used for slitting large sheets of metal.

Torsion An external stress which produces twisting within a body.

Trammel points Sharp points, usually mounted on a long bar, and used to transfer dimensions from one location to the other.

Transverse rows Layout of rivets in diagonal rows.

Two-piece fastener Bolt or rivet installed using a nut or a locking collar.

Universal head rivet A rivet design conforming to AN470 or MS20470 standard.

Upset head The shop formed head.

Variable-camber flap An auxiliary flight surface with a variable airfoil, or curve.

Warren truss A truss structure in which diagonal members carry both compressive and tensile loads.

Work-hardening Another term for strain-hardening.

Answers To Study Questions

Chapter I

1. Wings, fuselage, and empennage.
2. Ailerons, rudder, and elevators.
3. Spoilers, flaps, Krueger flaps, leading edge flaps, and trim tabs.
4. A wing with external bracing.
5. A fuselage with stressed skins and few internal braces.
6. Bulkhead.
7. Longeron.
8. Elevator, horizontal stabilizer, rudder, and vertical stabilizer.
9. An increase in bearing load due to compression.
10. Monocoque and semimonocoque.
11. They reduce lift on a wing in flight and during landings.
12. The braced wing has external support, and the cantilever wing has internal support.
13. Spars.
14. Ribs, stringers, and spars.
15. To provide maximum strength with less weight.

Chapter II

1. 140°.
2. 59°.
3. Pneumatic.
4. 1-80.
5. #30.
6. Cooling, lubricating and removing drillings.
7. Copper.
8. Universal and countersunk.
9. 3 to 5 LBS.
10. Tang, heel, face, edge, and point.
11. A file with two sets of teeth.
12. In the direction of rotation.
13. Router.
14. Add soft aluminum strips around the nose of the brake.
15. High.

Chapter III

1. Aluminum alloy.
2. Zinc.
3. 4.5%.
4. .12%.
5. Hardness.
6. Magnesium, aluminum, titanium, stainless steel, and iron.
7. Copper.
8. Magnesium.
9. Cold working.
10. Age hardening.
11. Heating, quenching and aging.
12. Solution heat treated, then cold worked.
13. 250°F.
14. Annealed.
15. 7XXX, 6XXX and some 2XXX alloys.
16. Alclad, 24, heat solution heat treated and cold worked.
17. Light, strong, and corrosion resistant.
18. Special fasteners, fire walls, and hot section skins (stainless steel).
19. 6% aluminum and 4% vanadium.
20. 4130.

Chapter IV

1. 2024T4, 2117T4, 2017T4, and 7050T73.
2. Silver, copper, black, and gold.
3. #30.
4. 7050T73.
5. 5056H32.
6. Dimple.
7. To temporarily remove age hardening.
8. 2017T4 crack-free.
9. 34,000 PSI.
10. DD.
11. Bearing or tension.
12. Driving rivets into a 60° or an 82° well.
13. Cold dimple countersunk rivets.
14. 4D.
15. $DSS = f_s \times A_s \times N$.

Chapter V

1. Cherrybucks, Lockbolts, and Taper-Loks.
2. Taper-Lok.
3. Shear and tension.
4. Because the center stem is locked in place with a ring.
5. Cherrymax, Olympic-Lok, and Huck-Clinch.
6. 1/8, 5/32, and 3/16 of an inch.
7. It must be 1/32 inch larger than the solid-shank rivet.
8. Huck NAS 1900.

Chapter V (cont'd.)

9. File off the retainer ring, tap the stem out, and drill the MFG head off.
10. Can be used on any head within the size range.
11. 6AL-4V upper shank and Ti-Cb lower shank.
12. Does not use a collar; it is driven like a solid-shank rivet.
13. Where solid-shank rivets cannot be bucked.
14. Countersunk and universal heads.
15. Dzus, Camloc, and Airloc.

Chapter VI

1. Radius, thickness, and number of degrees of bend.
2. BA increases.
3. BA decreases.
4. The amount of metal subtracted from the finished leg of a bend. Find the unbent portion: SB = (R + T).
5. A = .793; BA = .281; C = 3.586; BA = .281; C = .793.
6. A = .793; BA = .421; B = 1.543; T = 2.757.
7. BA = .286.
8. BA = .289.
9. All bends less than 90°.
10. Divide bend angle by 2 and use tangent table.
11. Temper and thickness of the metal.
12. Center line of the metal in edge view.
13. On the neutral line between bend radius lines.
14. Stretches.
15. Compresses.

Chapter VII

1. AC43.13-1, manufacturers maintenance manuals, and/or military TOs.
2. Monocoque skin repairs, corrugated skin repairs, and spar splices.
3. Analyze the damage, remove the damage, install new parts, and inspect the job.
4. To order new regulation parts.
5. Round, oval, and square with one-half inch radius corners.
6. Surface.
7. Cold working.
8. 40 rivets.
9. An authorized inspector (AI).
10. Two.
11. Owner, FAA and mechanic file.
12. RPI = T × 75,000 ÷ driven Shear Strength.
13. 22,032.
14. Under the wing near a spar or rib.

Chapter VIII

1. Intergranular.
2. Fretting.
3. Stress.
4. Interior surface corrosion.
5. Alclad, anodizing, or alodizing.
6. Solidifies aluminum oxides on the surface of the metal.
7. All oxides must be removed or solidified.
8. Alcladding.
9. It will corrode until it exfoliates.
10. Magnesium.

Chapter IX

1. A2024T3.
2. 4130 steel.
3. Aircraft manufacturer.
4. Light weight and strength.
5. Slip rollers.
6. Give shape to the leading edge.
7. A2024T4.
8. The spar-to-tab and two front rib tabs.
9. A wing.
10. To give a smooth finish and a rivet surface.
11. Seven.
12. Gives strength to the area.
13. Drive them according to the minimums list in AC43.13-1.
14. Slip rollers, bench shears and box pan brake.
15. Safety.

Index

D

E

F

G

H

S

T

U

V

W

X

Z